The Inner History of Devices

edited and with an introductory
essay by Sherry Turkle

The MIT Press Cambridge, Massachusetts London, England

For information about special quantity discounts, please email
special_sales@mitpress.mit.edu.

This book was set in Bookman Old Style, ITC Bookman, and Stymie
by Graphic Composition, Inc., Bogart, Georgia.

Printed and bound in the United States of America.

Library of Congress Cataloging-in-Publication Data

The inner history of devices / edited and with an introductory essay
  by Sherry Turkle.
      p.   cm.
    Includes bibliographical references and index.
    ISBN 978-0-262-20176-6 (hc : alk. paper)
    1. Technology—Psychological aspects. 2. Medical technology—
  Psychological aspects. 3. Computers—Psychological aspects.
  4. Internet—Psychological aspects. I. Turkle, Sherry.
  T14.5.I5643   2008
  303.48′3—dc22
                                                    2008005530

10 9 8 7 6 5 4 3 2 1

To Harriet Turkle and Mildred Bonowitz

Contents

Acknowledgments

In my teaching, I am often asked what stands behind an ethnography alert to the inner life of its subjects. When students press me for guidance, perhaps a list of questions, I tell them that intimate ethnography is not advanced by a particular agenda, but through creating a space to listen. The questions we ask in an intimate ethnography of devices—how do people feel about the objects in their lives? how do relationships form around them?—are common to many ethnographic traditions. Intimate ethnography explores the many ways that the first answers to these questions are not the last answers. It attends to untold stories. People have a received wisdom about themselves, a kind of "company line" about their lives. To get beyond these, it is helpful to infuse ethnography with other disciplines of self reflection—specifically, the sensibilities of the clinician and the memoirist. This book grew out of my efforts to answer my students' questions about methodology. I thank them for both inspiring and contributing to it.

My conversations with students about technology and the inner life have most recently taken place in the workshops, seminars, and classes of the MIT Initiative on Technology and Self. There, the close study of objects gives memoir and psychodynamic thinking a role in the training of social scientists. I founded the Initiative in 2001, and I thank the Mitchell Kapor Foundation for making it a reality, and the Kurzweil Foundation, the Spencer Foundation, the National Science Foundation, and the Intel Corporation for making it possible for it to continue its work.

This book is the third in a series that has grown out of work at the Initiative. I thank all who have participated in its programs, with a special debt to my students in the Initiative courses "Things and Thinking" and "Science, Technology, and Memoir." This volume draws on the research program funded by the Spencer Foundation on "Adolescence, Technology, and Identity" and that of the Intel Corporation on "Nurturant Technology" as well as on the hard work of several Initiative working groups, including Psychoanalysis and Digital Culture; Body Technology; and Information Societies, Technologies and Self. My academic department at MIT, the program in Science, Technology, and Society, has made a good home for all of these endeavors.

In addition, I thank those research associates at the Initiative who worked to shape this book: Olivia Dasté, Anita Say Chan, Robert Briscoe, and Will Taggart. Kelly Gray, as in other publication projects at the Initiative, was decisive in helping this book reach for better ideas as well as greater elegance and clarity. During the writing of this book, I craved the time and the tranquility that can only come from a smooth-running professional life. Here I thank Claire Baldwin, Grace Costa, Michele Crews, Trude Irons, and Judith Spitzer. Ms. Spitzer's eye for detail was a great friend to this book. At the MIT Press, special thanks to Deborah Cantor-Adams, Erin Hasley, Colleen Lanick, Alyssa Larose, and Robert Prior.

My daughter Rebecca has grown into young womanhood as this collection matured. Readers will be much helped by the consistency of her remarks as I worked at the kitchen table. Every time Rebecca said, "No one will know what that means," her academic mother looked long and hard at a particular phrase of well-loved jargon. Additionally, it must be said that Rebecca's proofreading skills are the stuff of legend. Thank you, Rebecca, for this and everything else.

My mother was about feelings and communication; my aunt was about ideas and taking commitments through to the end. Trying to bring their messages together inspires my personal and professional life every day. This book is dedicated to their memory.

Boston, Massachusetts
Spring 2008

Reading the Inner History of Devices

INNER HISTORY

Sherry Turkle

Thirty years ago, he was holding a TRS-80 home computer and I saw tears in his eyes. "This computer means everything to me," he said. "It's where I put my hope." I began the interview thinking I would learn something about how computer hobbyists were putting their new devices to work. By the end of the interview, my question had changed: What was there about personal computers that offered such deep connection? What did a computer have that offered hope?

Since then, studying people and technology, I have learned to listen attentively at such moments. The stories I hear usually have little to do with the stated purposes of the technology at hand:

"When I listen to my speech synthesizer, I hear it as an inner voice."

"I wasn't even sure I had sent that email, until I got your reply. I thought that maybe I had only dreamed sending that message, or fantasized it."

"Everything that I was interested in and everything that was important to me was on that Web site."

These three voices, all from this collection, have much in common. They refer to attachments in which technology inhabits the inner life and becomes charged with personal meaning. One voice is from a memoir, one from the clinical notebooks of a psychoanalyst, and the third from the field notes of an anthropologist, an ethnographer.[1] Without attribution, it would be hard to say which is which.

Here I bring together these three traditions—memoir, clinical practice, and fieldwork or ethnography—through which such voices emerge. Each tradition suggests a way of listening that adds new dimension to our understanding of how technologies affect our relationships and sensibilities. Each illuminates the subjective side of the technological experience, how what we have made is woven into our ways of seeing and being in the world. Together they enable us to read the inner history of devices.

Three Ways of Listening

In general, we treat the memoirist, clinician, and ethnographer as members of different tribes. The first we see as artists; the second we hold to the standards of healers (in America, until recently, psychoanalysts were required to be physicians); the third we call social scientists. We ask the first to make things that are beautiful, the second to be efficacious, and the third to be accurate.

These divisions, however real, can distract from important commonalities. Memoirists show us social forces lived out in personal experience. They do a kind of ethnography in the first person. The clinician has a close view of cultural pressures shared by those who are in treatment and those who are not. And ethnographers collect fragments of memoir from their subjects, approaching, in the words of anthropologist Clifford Geertz, large interpretation and abstract analysis "from the direction of exceedingly extended acquaintances with extremely small matters."[2] Indeed, in the 1970s, Geertz memorably described the ethnographer as practiced in the art of conversation.[3]

Geertz's notion that conversation is at the heart of the ethnographic encounter frames the field of anthropology as interpretive to its core, close to the reflections of memoir and the informed guesswork of the clinical life: "Cultural analysis is (or should be) guessing at meanings, assessing the guesses, and drawing explanatory conclusions from the better guesses, not discovering the Continent of Meaning and mapping out its bodiless landscape."[4] Indeed, in all three traditions, the work begins with conversation: a conversation with self, a conversation with an other. All three are disciplines of self-reflection. Together they change our understanding of our lives with technology.

When memoirists bring their artistic sensibilities to recollections of technological intimacies, new insights emerge. Similarly, clinicians discover untold stories when they explore their patients' relationships with technical objects—cars, motorbikes, stereos, and most recently computers and virtual worlds. When memoir and clinical sensibilities inform ethnography, they can shape a deep intelligence about technology and the inner life. Put otherwise, the inner history of devices calls for an intimate ethnography. Classical ethnographers are skilled listeners; intimate ethnographers, as the aphorism goes, listen with the third ear.

This kind of ethnographic work is hard to do because people find it difficult to talk about technology in ways that don't follow a standard script. We approach our technologies through a battery of advertising and media narratives; it is hard to think above the din. In contrast, the inner history of devices is about stories not heard unless one begins with quiet. Intimate ethnography takes patience; it makes room for people to discover what is really on their minds; it creates a space for self-reflection.

"All my life I've felt that there was something magical about people who could get into other people's minds and skin, who could take people like me out of ourselves and then back to ourselves."[5] This is memoirist Anne Lamott speaking about literature, but her sentiment could have been directed toward the psychoanalytic enterprise or the work of anthropologists, which begins with a displacement from their own cultures, in order to see it with fresh eyes.

In her essay on those who claim to be addicted to the Internet technical news site Slashdot.org, anthropologist Anita Say Chan immerses herself in a world of unrepentant addicts. This is her cultural displacement. Chan's informants are telling her something that seems to make little sense. It is not just that they are saying they are addicted and they like it. They are suggesting that in losing control of their behavior, they have come to a better place—politically, socially, and even economically. Chan's informants have reversed the social meaning of addiction. They struggle to describe why they are happy to be living under compulsion. As students, they are not going to classes; as employees, they are falling down on their jobs. Nevertheless, they insist that Slashdot enhances their lives. Among other things, it increases their political awareness and helps them appreciate their personal learning styles.

The Slashdot addicts are confirmed in their passion; they also find their vulnerability to the site frightening. In her conversations with them, Chan respects their positive feelings and their anxiety. She refrains from making the addicts "feel better" about their choices. (To keep the conversation going she does not say, "Oh well, you're not talking about 'real' addiction—your way of talking is just a turn of phrase.") That might relieve anxiety and lighten tension for a moment, but it would deny the self-described addicts a context to more fully reflect on their situation. Her informants have to live with ambivalence; Chan does as well.

Chan creates a safe space for contradictory feelings to coexist, making possible a conversation in which the standard meaning of a word like "addiction" can be called into question. Geertz said of anthropology that "coherence cannot be the major test of validity for a cultural description."[6] Nor can it validate a description of how technology enters the inner life. The ethnographic space has to generate its own kind of coherence. It is, in spirit, similar to the coherence that writers create around themselves, one that leaves room for complexity and contradiction.

Virginia Woolf calls the writer's space "a room of one's own." In its safety and containment, writers open themselves to the kind of reflection, where, as for Chan's addicts, feelings don't conform to a predetermined script. In turn, the psychoanalytic tradition needs a space that is "transitional," a space removed from everyday life; it is a liminal space, located between things.[7] There, relationships are not based on timeworn hierarchies but develop in new, meaning-filled encounters.[8] In analysis, transitional space facilitates understanding that can lead to change. The intimate ethnographer creates this kind of space, not to change the lives of informants but to illuminate their experience.

Transitional space is a metaphor that carefully undermines any view of ethnographers as spectators "soaking up" elements of a field setting. Everything about the sponge metaphor is wrong. Ethnography is not a passive practice; understanding the experience of others demands active listening. Nor are ethnographers trying to "get something" out of their subjects through clever questioning. They are trying to create an environment where what is there can emerge.

The Prepared Listener

In social sciences, there is an understandable emphasis on getting permission from subjects—obtaining informed consent to study them. But studying how technology enters the inner life also requires that you give permission to subjects. People feel permission to speak when they trust a researcher's promise of confidentiality and when they believe they are engaged in a process that will help them make better sense of their own experience. The motivation of the ethnographic subject is not simply to help the field researcher. It is increased self-knowledge. Ethnographers make it more likely that their subjects will achieve this by preparing themselves as listeners. In this, the psychoanalytic tradition has much to teach them, most centrally that effective listening begins with a measure of self-knowledge.

The central element of classical psychoanalytic training is a personal psychoanalysis.[9] The idea behind this practice is not simply that one learns "how to" by having one done "to you." Rather, self-knowledge is crucial to psychoanalytic practice because analysis does not proceed by providing information but through a shared experience in which the analyst is a catalyst for change. So, for example, it is expected that, in the course of therapy, the patient will develop feelings for the analyst that are the result of unresolved feelings from other relationships, feelings that are known as the transference. Recognizing and analyzing these feelings is a crucial motor in therapy. This is one way that issues of the patient's life are brought into the safe space of the consulting room, not by experiencing them in recollection, but by reliving them in the transference, where they can be analyzed. Transference does more than project the past; it reflects each individual's patterns of connection. As such, it can provide the energy to relate to new objects, people, and situations. Understood most richly, it is the power to invest in life "in one's own way."[10] Beyond the transference, it is also expected that the analyst will develop feelings for the patient, feelings referred to as the counter-transference. The analyst has to be able to understand and use these feelings for the analysis to progress.

In the analytic context, the privileged relationship between therapist and patient has been called a "therapeutic alliance." It is a working relationship in which both share a set of goals: the patient's greater self-understanding and, from there, a greater emotional range. The hope is that this increased range will translate into greater resiliency, capacity for joy, fulfillment in work and love. The method is based on the assumption that, with self-knowledge, we are better able to stand back from self-defeating patterns. To the question, "How does one stay open to complexity and levels of meaning that may surprise you?" the psychoanalytic tradition answers, through self-knowledge and new emotional practices. Only then, can one resist shutting down when material becomes threatening or confusing.

Ethnographers, too, need inner preparation. They, too, need to see how individuals invest in life "in one's own way." They, too, negotiate a complex range of feelings as they work. Close to people's intimate experience, they hear difficult things, things that touch close to the bone. Like psychoanalysts, ethnographers need to be present yet able to maintain distance. They need to tolerate ambivalence rather than intervene to make things seem more "coherent" or easy to accept. They need to facilitate conversation, yet maintain the boundary between listener and informant. They need to know which hesitations in conversation indicate deep feeling that should be pursued and they need to know when to stop. Ethnographers, like clinicians, form alliances with their subjects. Ethnographers share the goals of their particular enterprise with their informants, making sure they understand that the researcher is interested in such matters as why someone is attached to a white dialysis machine but dislikes a blue one. These are precisely the sort of matters that anthropologist Aslihan Sanal attends to in her study of dialysis patients awaiting kidney transplants.

Sanal offers her informants something that their attending medical personnel cannot. Oguz, a young Turkish patient, tells Sanal that the kidney specialists around him have an agenda—they want to move him toward a transplant. The psychiatrist sent to speak with him also has an agenda—to reduce resistance to the procedure. As this unfolds, Oguz finds the ethnographer to be the safest interlocutor: "A year ago, I made a suicide attempt and I was taken to the hospital," he says. "There I visited a psychiatrist, but I did not tell her anything. In dialysis, a psychologist approached me, but I did not say anything. Now I am talking to you."

Why did Oguz speak this way to Sanal? The answers recall how Chan came to hear stories from those who claimed Slashdot addiction. Like Chan, Sanal was nonjudgmental and made her informants feel safe. She moved into their world. She forged an alliance toward understanding. Her essay reminds us that particularly when one works with disadvantaged or ill people, with people one feels one could help, it is natural for ethnographers to have fantasies that they might rescue their informants. But just as clinicians master the countertransference in therapy—feelings toward and aspirations for patients that have the potential to interfere with therapeutic work—ethnographers cultivate respectful reticence. Their subjects are not there to be lectured or reformed. When ethnographers offer informants a space in which to know themselves better, they are offering a great deal.

When Oguz receives his father's kidney, he becomes depressed. He sees his new situation as worse than being in dialysis. Unguarded with Sanal, Oguz shares his revulsion at being this closely associated with his father, someone he has always disliked. Oguz begins to describe symptoms that point to his identification with his father. Formerly fastidious in his habits, now Oguz does not use soap, will not wash his hands or shave. He says: "I have not washed my hands for two days. Since the transplant I can no longer wash my hands. . . . If only I knew why . . . maybe I could start playing with soap again."

Oguz does not understand where his symptoms come from. Sanal has some ideas—she notes that Oguz is taking on his father's habits, but she shares this with the reader and not with Oguz. Although Sanal is not his therapist, she has borrowed certain tools of the clinician to help her sort through the complexity of this case. In Sanal's inner history of the dialysis machine, Oguz's language sometimes sounds like what he might say in a memoir (if he were writing one) and sometimes sounds like what he would say to a therapist (if he were seeing one).

Sanal's work underscores the complexity of relationships with cutting-edge technologies. When doctors save a life by beginning dialysis or performing a kidney transplant, most patients are grateful for the lifesaving procedure, but alongside these feelings there can also be distress—patients are coping with something radically new. In ambivalent relationships, many feelings coexist without negating each other; intimate ethnography is dedicated to hearing conflicting inner voices.

Psychoanalysis or dynamic psychotherapy (this is psychotherapy done in a psychoanalytic spirit, but not necessarily in the classical, four times a week, "on the couch" method) can be of great use to every ethnographer. There, one learns one's own limitations, one's vulnerabilities, when one is most likely to project one's own feelings onto others. One learns to be a more discerning listener and not to trust the first thing that is said. One learns to pay full attention and the difference between full attention and what usually passes for listening. One learns to let one's thoughts find new associations. What things mean is often hidden from us, accessible only by indirect routes. One learns to attend to dreams.

These days, the "talk-therapies" of the psychoanalytic tradition are rather out of fashion. As a way to "feel better," the ostensibly more speedy cognitive therapies and psychopharmacology have taken center stage. As a teacher of ethnography, I could argue for the didactic power of psychoanalysis or dynamic psychotherapy, but these are interventions that only an individual can choose, major financial and emotional investments.[11] Professionally, I am committed to something more realistic, something I can do in my classroom: asking ethnography students to read and write memoir. Through memoir, ethnographers learn about their own inner life and how to see the general in the particular. They are better able to hear when their informants struggle to do the same thing.

Studying people and their devices is, quite simply, a privileged way to study people. As William James reminds us, thought and feeling are unified in our apprehension of objects.[12] Technology serves as a Rorschach over a lifetime, a projective screen for our changing and emotionally charged commitments.[13]

Untold Stories

The concerns of inner history are not exotic. For example, one could be interested in life with everyday technologies and be satisfied with reasonable answers: people are vulnerable to "Internet addiction," patients implanted with internal cardiac defibrillators gain a new lease on life, cell phones enhance connection and communication. Each of these is, in its way, a "company line" on a new technology. And all of these may well be true as far as they go. The essays in this collection go further, offering moments when we learn something that breaks with conventional wisdom. At these moments of new truth, people express themselves in ways that are highly particular, close to idiosyncratic—and often the body is involved.

The anthropologist Anne Pollock studies patients who have been implanted with internal cardiac defibrillators, devices that will shock the heart if it fails. These patients know what they are supposed to feel about their implants. They are supposed to feel grateful. And they do. But they also suffer from their new cyborg status, something signaled by the distinctive diction they use to talk about being shocked and revived: "I died and then. . . ." The experience of receiving multiple, painful shocks, of never knowing when one will be shocked, or what actions will trigger a shock, leads ICD patients to develop rituals to prevent shocks and magical thinking about how they might be warded off.

The historian Michel Foucault wrote of prisoners who learn self-surveillance by internalizing the gaze of their prison guard.[14] Pollock notes that for ICD patients, surveillance "begins by being within." For them, there is no self that is independent of the device: patient and defibrillator are one.

Many ICD patients feel that the device has cheated them of the death they would want, a simple death from a heart attack. One patient is nostalgic for the death he almost had before he received his ICD: "Blacking out on the road, dying like that would be nothing. There would be no pain whatsoever." Yet ICD patients seem to agree that to remove the defibrillator would be the moral equivalent of suicide. A patient's wife says: "I don't think there's even that option. You cannot be the old you." Getting an ICD is, for her, nothing short of a "metamorphosis," like having children. Her husband agrees. "You can't get rid of kids either." Becoming cyborg is not a reversible step.

Alicia Kestrell Verlager, blind in one eye and going blind in her second, dreams of herself as cyborg. Few would suspect that this sightless student, dutifully using her computer as a reading machine, believes that prosthetics turn her into a repaired, if flawed, machine. Verlager's memoir about her prosthetic eyes illustrates one place the inner history of devices can take us—to people learning to write the story of their own lives. Like Pollock's subjects, Verlager is aware of the triumphalist narratives that technology offers. They would imply, for example, that medicine and its devices would make her eyes steadily improve. These narratives would put her doctor in the role of "the hero who saves my eyes and me as the brave and scrappy orphan who overcomes blindness."

> But I've been miscast, because really, I am already looking forward to the day when we can all acknowledge that my eyes are past saving, and I will no longer have to deal with needles and bright lights and hospitals and the frustrated anger of doctors and family members who cannot accept me as blind and in pain.

Verlager doesn't want to be at war with a part of herself that is weak and failing. She doesn't want to "fight" to get better. She prefers to think of her whole imperfect body in the same way she thinks of her grandmother's 1966 Ford Falcon. "I have a certain loving acceptance that its shortcomings are just part of what it is. I see my body as technology." She is ready to accept blindness. She has one prosthetic eye; she wants a pair. With the notion of technology as prosthetic, Verlager sees the possibility of stepping aside from her lifelong struggle with her biology. For her, posthumanism is not a theoretical position.[15] It is her nascent identity. Year by year, bit by bit, she adds on new "stuff." "I have come to think of all my electronic devices as prosthetics much like my eyes," she says.

> First there was my original prosthetic; then my adoption of computers and synthesized speech and the replacement of my second eye with another prosthetic; finally, I add the technological

"tethers" of a cell phone and digital recorder. With these final tethers, I began to visualize all information—speech, text, whatever—as 0s and 1s, which can be converted and catalogued in digital formats.

Verlager's speech synthesizer becomes an "inner voice," and she finds herself using and forgetting the computers that have become pieces of herself. She says of them that they "have blurred the boundary between me and not me. I sometimes think of myself as becoming science fiction."

Other medical technologies induce similar fantasies. Pollock's ICD patients come to think of themselves and their machines as one. Sanal's dialysis patient Oguz sees himself as "half robot and half human," an electronic "thing" that has "exited the human condition." Zehra, another dialysis patient, feels that her body is no longer her own, that it is being replaced in the cycles of dialysis. During dialysis Zehra tries to sleep but her dreams are mostly recursive, dreams of herself on dialysis. Sanal remarks that "connection to the dialysis machine eroded the boundaries of the ego."

In Natasha Schüll's essay on compulsive gambling, we learn that the gambler's goal is not to win, but to stay connected to the machine. Indeed, one of the gamblers, Julie, confesses that winning can be annoying: "If it's a moderate day—*win, lose, win, lose*—you keep the same pace. But if you win big, it can prevent you from staying in the zone."

Another gambler, Isabella, associates her absorption in gambling machines with a televised science fiction program in which characters are sucked into computer screens and from there into computer games. When Isabella plays video poker she feels "inside the machine, in the king and queen turning over, almost hypnotized into *being* that machine." Images of cyborg coupling run through the community of gamblers. One of them, an electronics technician named Randall, says:

> I get to the point where I no longer realize that my hand is touching the machine, I don't feel it there. I feel connected to the machine when I play, like it's an extension of me, as if physically you couldn't separate me from the machine.

Voice synthesizers, dialysis machines, gambling machines designed to rivet their users—these are, from the start, devices within or close to the body. But everyday communications technologies—such as cell phones—these too become intimate machines that inspire strong, even eroticized attachments. So while public discussions of cell phones stress their convenience and how they make our lives more productive, E. Cabell Hankinson Gathman's memoir takes the narrative in a very different direction.

As a young woman in the final throes of a love affair, Gathman is surprised to find her cell phone a crucial actor in her drama. She sleeps with the phone in her arms, waiting for the ring tone she has set to herald her lover's calls. When they break up, she deletes his phone number within hours but cannot bring herself to delete the special ring tone she programmed to signal his calls. Long after the relationship is over, his ring tone remains. Of course, it is no longer activated by her lover's calls—he no longer calls.

Gathman leaves the ring tone on her phone to be chanced upon:

> Whenever I cycled through my store of tones, to change my default or set a tone for a particular contact, there it was. . . . Every time I heard it, it shocked me for a moment with instinctive pleasure, the sense that we shared something, that he wanted to talk to me, before it filtered through

the present and I remembered that he hadn't even called to say he wasn't going to call anymore.

For Gathman, the ring tone becomes part of a mourning process. Others engage in similar rites: for many, deciding how to deal with the voicemail and greeting messages of a deceased family member has become part of the process of saying goodbye. Some people keep these tones and messages but find that having them is more important than listening to them. Erasing them seems intolerable. For others, deleting these messages is part of the ritual of mourning.[16]

A moment of mourning is at the heart of Rachel Prentice's essay on digitized bodies. An anthropologist, Prentice studies the Visible Human Project, a computer program that digitally reconstructs a cadaver. Prentice tells us that its images—one of a man, one of a woman—"resemble photographs of people, but the images are strange. Color and shadow are subtly wrong. There are odd marks. They seem to have no context other than the computer screen."[17] To get these images, the bodies had to be specially embalmed, then frozen, scanned, and sectioned into microscopic slices. Then, digital photographs were taken of each shaved-off body section. After it all, the frozen sections were cremated.

Prentice describes an interview with Julie, a forty-six-year-old medical writer. Julie's first response to computer images of the "visible woman" is dispassionate. She shows more interest in the download speed of the program than in the images themselves and claims that she has never seen a dead body. But as Julie focuses on the dead woman's images, her response changes, "The eyes look painfully closed. There's a furrowed brow." Julie imagines that she sees pain on the woman's face, pain caused by something that happened as her body was being manipulated on its way to becoming a program. As the interview progresses, Julie decides that the violence to the woman occurred before her funeral, "in the casket, before it was decided she would become the visible woman." Julie assumes there was a funeral with a casket.

The turning point in Julie's interview comes after the formal interview is over. Prentice tells us that "at the end of the interview, after I have closed my notebook," Julie, who only a short time ago said she had never seen a dead body, now remembers that she has seen three of her grandparents in death. We owe this powerful second interview to the evocative nature of the Visible Human Project and to the quality of the moment that the ethnographer has created with her subject. Julie begins to talk about a traumatic incident involving her grandmother's body in her casket, the place that she has put the visible woman at the moment of her pain. At the funeral home, Julie noticed that her grandmother's lipstick was awry. "Improperly applied lipstick would have upset her grandmother," says Prentice. "Though she knew that her grandmother was dead, the improperly applied lipstick forced Julie to confront all that attends death: the body lying in front of her no longer had agency. It was at the mercy of strangers," professional attendants.

Julie's associations went from the discomforting exposure of the visible woman to her grandmother's vulnerability in death. Prentice knows, as she is closing her notebook, that another kind of interview has begun. Julie began her interview with Prentice as a medical writer. She ends it as a granddaughter.

There is an instinctive desire to protect the humanness of those who are vulnerable. The visible woman is shown to us past our ability to protect her. She was turned into a thing that is allowed no modesty. Prentice's intimate ethnography did not lead to a criticism of technology but to an implicit question about consent: did the human being who became the visible woman really know what was to become of her body, what it would mean to be turned into a program? She is a human being now used in a new way, splayed out in a virtual world where she is at the disposal of anyone who cares to look. The classical anatomy laboratory is a highly ritualized space. Those who cross its threshold are made to qualify. Bodies on the Internet are just another window on the screen.

Prentice begins her essay by describing her informants as socialized into professional roles that prepare them to view computer images with dispassion—the role from which Julie departed. The meanings of technology are constructed in culture, or as Geertz puts it, "Culture is public because meaning is."[18] Certain styles of technological attachment become dominant in particular places and times; examining individual relationships with technology can be a window onto larger social forces.

So for example, Schüll's consideration of video poker addiction leads her to reflect on the 1973 comment by sociologist Daniel Bell that, as individuals moved away from the assembly line and joined the service economy, they would talk to each other more.[19] Thirty-five years later, Schüll reflects that Bell's theory did not anticipate that today, talking to other people is somehow experienced as exhausting; we happily retreat into worlds where we are content to communicate with machines. Schüll describes gamblers with a pathology, but her account raises a question that confronts us all: what is there about modern community that has so taxed us that we sometimes prefer to reconstitute it through machines?[20]

Orit Kuritsky-Fox's memoir tells another story of technology, modern community, and retreat—her recollections reveal television as a window onto social divisions in Israeli society. For secular Israelis, television is a force of social cohesion. When Kuritsky-Fox was growing up, the country was small and the stations few. She describes a television program from her youth in which people moved through mirrors to rooms on the other side—a program so well known to her generation that twenty years after its broadcast, she can still hold a conversation about it with friends and colleagues. Yet, television also divides. Some sects of orthodox Judaism prohibit television. For the orthodox, admitting that one watches television puts one beyond the pale. In Israel, the society's secular/religious split is played out in a relationship to television. As secular Jews, Kuritsky-Fox and her immediate family watch television. Her grandmother lives with greater contradiction. She is "officially" orthodox; she can be buried in a sanctified cemetery only if television is not a part of her life. Yet, this woman's greatest pleasure is to chat in Yiddish to the TV news anchors as though they were old friends and offer family advice to the characters on soap operas—*Dynasty* and *The Bold and the Beautiful.*

So, when her grandmother dies, both she and her granddaughter have to "pass" as nontelevision watchers, one to be buried, the other to attend the orthodox funeral. Kuritsky-Fox, a television producer, describes her efforts to dress in the style of someone who would never think of watching television. She chooses a long dress, a dowdy look. She knows she should wear pantyhose in summer to complete the effect, but it is a hot day and she dispenses with the hose, something she thinks she can get away with. At the cemetery, Kuritsky-Fox comes to regret omitting the pantyhose.

> As I was waiting at the cemetery, the receptionist approached me and asked what I was looking for. I told him that I was there for my grandmother's funeral. He stared at me with dismay. I thought to myself that perhaps the pantyhose had not been dispensable. "Are you religious?" he asked, and I started to mumble. Then, he found a way to cut to the chase: "Do you own a television?" I hesitated for a second and then told him the truth. He gasped, looked into my eyes and asked his next question: "Your grandmother, she didn't watch television, did she?"

The cemetery scene captures the daily identity negotiations in this divided society, here crystallized around a communications technology. There is Kuritsky-Fox's grandmother, television her heart's delight, but whose aversion to television is certified by seven busloads of yeshiva students dressed in black suits that her orthodox sons have sent in as official mourners. There is Kuritsky-Fox's mother, an ardent secularist, for whom lying about television watching would be a betrayal. And there is Kuritsky-Fox herself, who has chosen to live in many cultures, a life resonant with the television program of her youth, in which people go through mirrors to the many rooms on the other side of them.

In the Verlager, Gathman, and Kuritsky-Fox memoirs, technology is central to forging identity, a central theme of this collection's clinical writings, which focus on adolescence and online life. These days, adolescents use life on the screen—social networking, game avatars, personal Web pages, and citizenship in virtual communities—to crystallize identity by imagining the selves they wish to be.[21] An online avatar can come to feel continuous with the self and so offer the possibility of personal transformation.[22]

Psychoanalyst Erik Erikson writes of adolescence as a time of moratorium.[23] Although the term implies a "time out," what Erikson has in mind is not withdrawal. On the contrary, the adolescent moratorium is a time of passionate experimentation, of intense interaction with people and ideas. The moratorium is not on significant experiences but on their consequences. Erikson writes that "the playing adult steps sideward into another reality; the playing child advances forward into new stages of mastery."[24] Time in cyberspace reshapes the notion of the moratorium because it may now exist as ongoing activity.

Yet this aspect of online life is scarcely recognized in most narratives about teenagers and the Internet.

Psychiatrist John Hamilton notes that no matter how serious their child's situation, parents of his young male patients come in with another complaint: "I can't get him off the computer." His clinical colleagues share these parents' negative bias, routinely advising them not to let their children on the Internet or, as Hamilton puts it, "tell[ing] them to get a real friend." Hamilton sees things differently. He uses therapy sessions to bolster adolescents' fragile identities.[25]

> In my practice I find that bringing the Internet into the therapy session enables difficult things to be said. The Internet takes a therapist and patient struggling to communicate only with words and offers them color, sound, and mobile avatars. The endless variety of Internet sites makes it possible for young men to find particular places and games that help them work on their inner life. They are even able to find characters to "play" that help them address specific psychological issues.

Many of Hamilton's young male patients have weak or absent fathers. They use the Internet to play the part of masculine superheroes. These superheroes are compelling because they offer images of strong men who do not need the attention of others. Hamilton is able to get his patients to reflect on the costs of their identification with these hypermales. To succeed in the "real" world, his patients need skills the superheroes lack— how to collaborate, share experiences, and understand others. Hamilton describes a patient who learns to question the simple equation of masculinity with aggression and lack of communication by "problem-solv[ing] for avatars in the virtual and tak[ing] what he learns back into the real."

The Internet is also a working material for psychoanalyst Kimberlyn Leary. One of her patients, a talented young woman named Morgan, needs to both criticize

Leary and be reassured that she is there, undamaged, to help her. When Leary goes on a business trip, Morgan, who experiences life in terms of limited resources, is enraged. If Leary has more, she will have less. Morgan writes a hostile email:

> Dr. Leary, my capricious shrink. You are the lamest thing going. The only thing that matters to you is your pathetic little writing endeavors. . . . You have no trouble clasping my check in those perfectly slender and inconspicuously manicured talons of yours without having been around for the most needful moments in my recent, unmomentous life. Right now, you are a pale substitute for my symptoms.

Leary responds to the email by focusing on how well Morgan has expressed herself and makes it clear that she is looking forward to talking more, that Morgan's feelings have not shut down their connection. And Leary sees possibility in Morgan's use of the phrase "right now"; it suggests that Morgan's feelings can change. In their next session, therapist and patient work to create a bridge between what Morgan could express in her email (which she says she has forgotten) and what she is able to discuss in person. The analyst and patient return to the image of Leary's "manicured talons," and Morgan admits that she has "always liked" Leary's fingers. "They are so long. I've always thought you probably played the piano and could make beautiful music with those fingers of yours." Email becomes a transitional space that opens new possibilities for communication.

In these clinical accounts, we are far from any standard view of Internet "addiction." The Internet appears as a medium in which people discover things about themselves, good and bad, usually complicated and hard to sort out. For too long, clinicians dismissed their patients' interest in email and in building virtual identities—on Web sites and computer games—as "just fantasy," as though that were not central to the business of being a therapist. But life on the screen can be working material for psychotherapy, not something that therapists should discourage as a waste of time.

There is a moment in psychoanalyst Marsha H. Levy-Warren's essay that illustrates how simple and profound it is to move beyond standard narratives. A patient, Joanie, an unhappy and overweight teenager, spends many hours a day playing a computer fantasy game, often instead of doing her homework. When Levy-Warren asks her why she finds her game so compelling, Joanie's first answer is the formulaic, "I know I shouldn't do it, but I just can't resist." This is the answer that has led to the many thousands of articles on computer "addiction." But Levy-Warren pursues the question, "Any sense of what is so irresistible about the game?" And she begins to get another kind of answer. Joanie says: "I really like who I am in it. You know, I created a character. It's a fantasy game." An inner history has begun.

By talking about what she can be on the game and cannot be in the rest of life, Joanie is able to consider how she has used her weight to remove herself from competition for male attention. She finds a way to acknowledge her own competitive feelings and her desire to be like her screen character—bold and assertive. "Gradually, her depression lifts," says Levy-Warren. "She feels closer in the real to the person she plays in the game and plays the game less." In psychoanalyst Christopher Bollas's terms, Joanie's game is revealed as an "adamant quest for a transformative object: a new partner, a different form of work."[26]

Collection and Recollection

The essayists in this collection consider devices that come supplied with sanctioned ways of understanding them. The authors take time to go further, often not knowing what they are looking for. In Nicholas A. Knouf's memoir, a medical device, in its presence and absence, allows him to dream.

Faced with his sister Robin's illness—the gravely debilitating Rett Syndrome that offers no hope, Knouf and his family learn of a technique that offers some. The family moves Robin's limbs rhythmically in crawling and walking movements on a specially built "patterning table." While Robin lives, Knouf is immersed in the community around the table. After Robin dies, the table's absence opens a reflective space for Knouf to consider how it shaped his life. "With Robin," says Knouf, "the volunteers, the discredited therapy method, and the patterning table, we had tried to awaken cognition with care, with the soft sheepskin, the men and women gathered around." But the table has stood in the way of many things: the animating force of Knouf's school work has always been the dream of curing his sister's illness; it takes him a while to find his identity after she and the table are gone.

With loss, Knouf is able to rework the past. Meaning, as Lillian Hellman wrote in her memoir, comes in *pentimento,* in the painter's layering of paint, in his "repentance" as he finds what he wants in the process of repainting. The meaning for Hellman is in "what was there for me once, what is there for me now," or as Geertz stressed, in the act of interpretation and reinterpretation.[27] What will become of this kind of reworking when, in digital culture, people's fantasies shift from telling the story of a life to having a complete record of it?

Computer and Internet pioneer Gordon Bell has immersed himself in the project of creating a full digital life archive. In 1998, he began the process of scanning books, cards, letters, memos, posters, and photographs—even the logos of his coffee mugs and T-shirts— into a digital archive. He then moved on to movies, videotaped lectures, and voice recordings. Faced with the question of how to organize and retrieve this mass of data, Bell began to work with a team from Microsoft.

The MyLifeBits project was born. Bell and his colleague Jim Gemmel describe the process of data collection:

> The system records his [Bell's] telephone calls and the programs playing on radio and television. When he is working at his PC, MyLifeBits automatically stores a copy of every Web page he visits and a transcript of every instant message he sends or receives. It also records the files he opens, the songs he plays and the searches he performs. The system even monitors which windows are in the foreground of his screen at any time and how much mouse and keyboard activity is going on. . . .
>
> To obtain a visual record of his day, Bell wears the SenseCam, a camera developed by Microsoft Research that automatically takes pictures when its sensors indicate that the user might want a photograph. For example, if the SenseCam's passive infrared sensor detects a warm body nearby, it photographs the person. If the light level changes significantly—a sign that the user has probably moved in or out of a room and entered a new setting—the camera takes another snapshot.[28]

What compels the architects of this program is the idea of a complete, digitally accessible life. To be sure, there are medical applications ("your physician would have access to a detailed, ongoing health record, and you would no longer have to rack your brain to answer questions such as 'When did you first feel this way?'"), but most of all, the authors speak of posterity, of MyLifeBits as a way for people to "tell their life stories to their descendants."[29] But what is it that future generations want to know of our lives?

In the collection *Evocative Objects,* architect Susan Yee describes her visit to the Le Corbusier archive in Paris on the day its materials were digitized.[30]

Yee began her relationship to Le Corbusier through the physicality of his drawings. The master's original blueprints, sketches, and plans were brought to her in long metal boxes. Le Corbusier's handwritten notes in the margins of his sketches, the traces of his fingerprints, the smudges, the dirt—Yee was thrilled by all of these. One morning, Yee has all of this in her hands, but by the afternoon, she has only digital materials to work with. Yee experiences a loss of connection to Le Corbusier: "It made the drawings feel anonymous," she says. More important, Yee says that the digitized archives made her feel anonymous.

When working in the physical archive, Yee was on a kind of pilgrimage. She did not pause in her work, so completely was she immersed in the touch and feel of Le Corbusier's artifacts. But once the material was on the screen, there was a disconnect. Yee found herself switching screens, moving from the Le Corbusier materials to check her email back at MIT. More than a resource, the digitized archive becomes a state of mind.

MyLifeBits is the ultimate tool for data collection. But what of recollection in the fully archived life? Speaking of photography, Susan Sontag writes that "travel becomes a strategy for accumulating photographs."[31] In digital culture, does life become a strategy for establishing an archive?[32] When we know that everything in our lives is captured, will we begin to live the life we hope to have archived?

The fantasy of a complete record for all time—a kind of immortality—is part of the seduction of digital capture. But memoir, clinical writing, and ethnography are not only about capturing events but about remembering and forgetting, choice and interpretation. The complete digital archive gives equal weight to every person, every change of venue. The digital archive follows chronologies and categories. The human act of remembrance puts events into shifting camps of meaning. When Bell and Gemmell consider the quantity of information on MyLifeBits, they talk about the "pesky problem of photograph labeling."[33] The program is going to use face-recognition technology to label most photographs automatically. In reading this, I recall childhood times with my mother in which she wrote funny things, silly poems, or sentimental inscriptions on the backs of family photographs. She liked putting them in a big drawer, so that, in a way, picking a photo out of the drawer, almost at random, was finding a surprise. Moments around the photograph drawer were moments of recollection in laughter, regret, sometimes mourning. Now automated for a steady stream of photographs over a lifetime, photograph labeling is just a technical problem. Bell and Gemmell sum it up by saying that "most of us do not want to be the librarians of our digital archives—we want the computer to be the librarian!"[34] In this new context, reviewing your life becomes managing the past. Subtly, attitudes toward one's own life shift; my mother, happily annotating her messy drawer of snapshots, never saw herself as a librarian.

Of course, the digital archive is only a resource; it remains for us to take its materials as the basis for the deeply felt enterprise of recollection. But one wonders if the mere fact of the archive will not make us feel that the job is already done.

Let me return to 1977 and to the man whose words began this essay, a middle-aged man with a TRS-80 home computer. This was Barry, living in a Boston suburb.[35] Barry knows how his TRS-80 works, down to every circuit. When typing a document, he imagines each instruction to the word processing program translated into assembly language and from there to the basic electronics of gates and switches. Whenever possible, Barry programs in assembly language to stay close to the bare machine.

Barry went to college for two years, hoping to be an engineer, then dropped out and went to technical school. He has a job calibrating and repairing electronic equipment for a large research laboratory. He likes his job because it gives him a chance to work "on a lot of different machines." But he comes to it with a sense of having failed, of not being "analytical or theoretical": "I always had a great deal of difficulty with mathematics in college, which is why I never became an engineer. I just could not seem to discipline my mind enough to break mathematics down to its component parts, and then put it all together."

Five years before I met him, Barry bought a programmable calculator and started "fooling around with it and with numbers the way I had never been able to fool around before." He says that "it seemed natural to start to work with computers as soon as I could." To hear him tell it, numbers stopped being "theoretical"; they became concrete, practical, and playful. Barry says that with the computer and calculator, "The numbers are in your fingers. . . . They put mathematics in my hands and I'm good with my hands." For Barry, what is important about having a computer at home is not what the computer might do, but how it makes him feel.

But what of his tears?

Ever since dropping out of college, Barry has seen himself as limited, constrained. Working with the computer has made him reconsider himself:

> I really couldn't tell you what sort of things I'm going to be doing with my computer in six months. It used to be that I could tell you exactly what I would be thinking about in six months. But the thing with this, with the computer, is that the deeper you get into it, there is no way an individual can say what he'll be thinking in six months, what I'm going to be doing. But I honestly feel that it's going to be great. And that's one hell of a thing.

It was at this point in the interview that I saw tears. Barry's world had always been divided into people who think they know what they'll be doing in six months and people who don't. Barry has crossed this line, and now he has started to call other lines into question, ones that have limited his sense of possibility. In school, his inability to do the kind of mathematics he respected made him lose respect for himself. The calculator and computer gave him mathematics. But more important than the mathematics he has mastered, he has come to see himself as a learner.

There are many stories to tell about people and their devices. We need to hear stories that examine political, economic, and social institutions. Inner history tells other stories. Inner history makes Barry's tears part of an ethnography of the personal computer. Inner history shows technology to be as much an architect of our intimacies as our solitudes. Through it, we see beyond everyday understanding to untold stories about our attachments to objects. We are given a clearer view of how technology touches on the ethical compacts we make with each other, compacts that philosopher Emmanuel Levinas has suggested begin when we look into the face of another human being.[36] Each essay in this collection brings us to the question we must ask of every device—does it serve our human purposes?—a question that causes us to reconsider what these are.

Through Memoir

THE PROSTHETIC EYE

Alicia Kestrell Verlager

My fourth-grade teacher, Mrs. McGuirk, once asked my class to write our autobiographies. I knew that I had no intention of telling my life story. As a half-blind orphan, I was uncomfortably aware that people reacted to my life story as if they were reading a Charles Dickens novel. I loathe Dickens novels. Not that Dickens didn't write some wonderful ghost stories, but his novels, those morality tales full of pathos and pity—what kid wants to be a character in one of those? Sentimentality is a powerful social weapon: the sentimental person can both pity and pry, but as the frequent object of such emotion, I often felt like the victim of a highway accident, penned in by a sea of rubberneckers.

This feeling of being gawked at by a crowd has left me with a kind of double vision. I often experience my life in the second or third person: I sense an audience experiencing me. "You're an orphan! How sad." "You're totally blind? How courageous." "You had arthritis as a baby? How strange." As a child, a direct question about my life would make me cringe. Now, I try to tell my own story and shape my audience's reactions. So, here I attempt memoir in several genres. I begin with a Coyote story. Coyote being Coyote, I don't think he would mind that my story changes shape halfway through.

Coyote is a Native American trickster figure. There are a lot of stories about Coyote, and most of them involve shape-shifting. One of my favorites is about how Coyote lost his eyes.

As Coyote was walking through the forest one morning, he heard someone say: "I throw you up and you come down in!" Infamous for his curiosity, Coyote followed the sound of the voice. He crept closer until, peering through some shrub, he saw Chickadee. Chickadee was throwing his eyes up in the air while saying, "I throw you up and you come down in!" and sure enough, his eyeballs would go up in the air and drop right down into his eye sockets, plop! plop! But when Chickadee saw Coyote, he got scared and ran away.

Coyote thought juggling eyeballs looked like fun, so he tossed his own eyeballs up in the air and said, "I throw you up and you come down in!" and sure enough, his eyeballs plunked right back into his eye sockets. Coyote did it again and again.

Somewhere between the tenth and hundredth time that Coyote was juggling his eyeballs, two ravens flying overhead spied him and thought it would be fun to play a trick on the trickster. Swooping down, each raven snatched one of Coyote's eyeballs out of the air and flew away with it.

Coyote cursed and kicked things for quite a long time, but then he decided to go after those ravens to get his eyeballs back. He had heard the ravens fly away and began to stumble in that direction, bumping into trees and tripping over roots and logs. Finally, after tricking a number of other animals out of their eyeballs and causing all sorts of trouble in these appropriations, Coyote got his own eyeballs back. As a person with two prosthetic eyes, I identify with Coyote in this story. Like Coyote, I find that prosthetics enable one to play with the boundary between self and not self, with what one can see and not see.[1]

If, as we often say, eyes are the "windows of the soul," I was afraid anyone staring into mine would only see broken ones with a dark ruin beyond them. As I think about this, I wonder if the preoccupation with windows is the reason people still call prosthetic eyes "glass eyes," since for decades they have been made from plastic. When people find out I have prosthetics, they inevitably ask, "You mean, glass eyes?" and I reply, "They're actually made of a lightweight yet durable polymer," an odd phrase that has stuck with me from the pamphlet I was given after my first eye surgery.

We look into someone's eyes and believe we see truth; we think of the eye as a direct link with another mind. In my own case, what is direct and what is veiled is complex because I see my prosthetic eyes as more "me" and more "real" than my "real" eyes ever were. Indeed, my "fake" eye looks more like other people's eyes than my "real" eye ever did. So the first time I wore my prosthetics at fourteen, there was relief from pain and the sense of engaging in a performance that was a pretense at its very core.

Through a Child's Eye

As a child, I was the author of a con worthy of Coyote. This was to persuade the adults around me that I was well adapted, that I could see more than I could, and most of all, that I was not going blind. I pretended to be a child rather than the old person I felt myself to be. To pull all of this off meant learning many little different cons. The first was learning how to fake seeing the eye chart. From when I was three until I was twenty, I read an eye chart at least once a month and sometimes once a week during visits to my ophthalmologist, Dr. Kassoff. The memories of these visits come with intense physical sensations that bring the past into the present.

It's useless to hate an inanimate object, but I hate the eye chart. Like the blackboards I stare at during school, I tell adults I can't see them, but the adults only reply, "Try harder." I sit in Dr. Kassoff's huge black leather chair and try to see. When that fails, he says, "Try harder," and I move to the performance of seeing. I learn that the way to do this is not to hesitate, to give the appearance of it being easy, effortless, the way it is for other people whose eyes barely glance at the chart before flicking away. I get good at looking for clues that tell me in which direction the E is pointing its fingers, realizing the corners are different for each direction, so I mostly stare at those. I wonder if this is what vision is for other people, a series of tricks that they know and I don't. I keep thinking that if I can learn the system or discipline of seeing, I will see better.

In fact, most of my eye problems stem from a painful form of glaucoma, acute angle closure glaucoma. The glaucoma makes my eye pressure much higher than most people's. The high eye pressure slowly destroys the optic nerve, which means that once vision is gone, it is gone, and is not correctable with glasses. As a child, one of the annoying questions I answer again and again is why I don't get better glasses. I become frustrated explaining to adults what the phrase "noncorrectable vision" means.

When Doctor Kassoff takes my eye pressure, I put my chin on a bit of plastic with a paper liner, and he extends a blue glowing light toward my good eye (he isn't interested in my bad eye because it is already blind and can't be saved). Due to its high eye pressure, my "good" eye is very light sensitive and feels permanently bruised. The pressure gauge hurts it. If I flinch, Dr. Kassoff gets angry and tells me I am wasting his time. My grandmother sits in the corner and tells me not to be a baby. I tell myself I don't feel pain. I make up a story in which my real self is a bird that flies away; even eye doctors staring straight into your eyes with lights and magnifying lenses can't see inside to the place where the bird flies.

In February of my thirteenth year, a glaucoma attack keeps me in constant pain. For the rest of my freshman year, I drag myself to school, although all I can do is lay my head on my desk and fold my arms around my head to block out the light. None of my teachers ask why I am in school instead of home, sick. In August, I get my first prosthetic eye.

Most people need scripts to guide them. There is no script for a kid who is going blind. All there is to work from is the drama in which blindness can be "overcome." So we play out this script. It casts my doctor as the hero who saves my eyes and me as the brave and scrappy orphan who overcomes blindness. But I've been miscast, because really, I am already looking forward to the day when we can all acknowledge that my eyes are past saving, and I will no longer have to deal with needles and bright lights and hospitals and the frustrated anger of doctors and family members who cannot accept me as blind and in pain.

Blindness (or "losing my eyesight," as most people prefer to say it—people really hate even using the word "blind") is always treated as the villain in this morality play. My eyes are the traitors, "weak" and "failing." I am supposed to be at war with the parts of me that are "bad" or "weak," actively "fighting to get better." Instead, I come to think of my imperfect body as I think of my grandmother's 1966 Ford Falcon. I have a certain loving acceptance that its shortcomings are just part of what it is. I see my body as technology.

Literary theorist N. Katherine Hayles writes of a posthuman view of the body that takes it to be "the original prosthesis we all learn to manipulate, so that extending or replacing the body with other prostheses becomes a continuation of a process that began before we were born."[2] I came to this worldview bit by bit. First there was my original prosthetic; then my adoption of computers and synthesized speech and the replacement of my second eye with another prosthetic; finally, I add the technological "tethers" of a cell phone and digital recorder.[3] With these final tethers, I began to visualize all information—speech, text, whatever—as 0s and 1s, which can be converted and catalogued in digital formats. I have come to think of all my electronic devices as prosthetics much like my eyes. When I listen to my speech synthesizer, I hear it as an inner voice.

I simultaneously use and forget my computers. They have blurred the boundary between me and not me. I sometimes think of myself as becoming science fiction.

The Prosthetic Aesthetic

My favorite *Star Trek* episode, "Is There in Truth No Beauty?" features a blind character, Dr. Miranda Jones. No one suspects her blindness because she wears a dress with an embedded sensor net. The sensor feeds her information about her environment and, at one point when Captain Kirk questions her abilities, she challenges him to compete with her at tennis, made confident by how much her sensor net has become part of her. As I began to think of my prosthetics as me, I saw them as my net, constantly evolving, always being tweaked, experimented with, tweaked again, and even sloughed off for new versions.

This is how I came to change my eye color when I got my second pair of prosthetic eyes. At that moment, my ocularist, the technician who manufactures my prosthetic eyes, is shocked and confused by my request because if I change eye color, "people will know" that my eyes are not real. I counter that for two months after the surgery that removed my eyes, my friends had seen me going about like the Phantom of the Opera with a bandage over half my face, and I am pretty sure most of them knew that something was up. My friends and I often joke about the bizarre eyes I could get: glowing eyes with slitted cat pupils, for example, or perhaps mirrored lenses? So I feel my request to the ocularist for a simple eye color change, something plenty of "normal" people do with colored lenses, is not irrational. What I get from this experience, however, is the subtle message that "passing," by preserving the original appearance of my body, is supposed to be more important to me than a desire to change my body.

Our culture holds two images of the body: in one, the body is static, if aging; in the other, individuals can re-shape their bodies, use them as a means of identity and self-expression. Users of prosthetics find themselves caught between the two images. Is their device aimed at restoring the static body or is it part of a continuous process of becoming? Is it better to "pass" for authentic or is it better to have the "unnatural" prosthetic one wants? Is a plastic, flesh-colored false leg better than a technically more advanced titanium-jointed one? The first can wear stockings and look more natural at a cocktail party; the second might enable its user to run.

While my own prosthetics have no ability to provide visual information, I refer to them as "my eyes," and to what I do with them as "looking." I think of "looking" as a turning toward, a gesture rather than an information-gathering process. The "turning toward" gesture is one of our most human. It is an invitation to communicate. When we do not wish to talk to someone we turn away, refuse to allow our eyes to engage, until the person demands, "Look at me!" To look is to allow ourselves to see, and to see is to allow something to be taken into ourselves.

For me, "rehabilitation" has meant dealing with other people who want to tell me what I should be allowed to take in, people who want me to self-censor. When I was learning to use a white cane, my rehabilitation instructor told me I was using it "wrong." I was using the cane to explore my surroundings, taking soundings of garbage cans, newspaper dispensers, and bike racks in an attempt to map my landscape. The instructor's method was designed to lay out a straight path and to use the white cane to signal to others that I was blind.

The Story Nobody Will Tell You

Before the first week of my sophomore year of high school, while other kids' back-to-school shopping revolves around new jeans and backpacks, my attention is focused on whether I will receive my new prosthetic eye in time for school. It is a close call: I spend the first day of school getting my eye fitted. On day two, I get on the school bus wearing the eye, but, while riding on the bus, I feel it slip out. I try a sleight-of-hand. I put my hand up as if to push my hair out of my eyes, then I let the hair fall back over my left eye. When I move my hand away, I have my prosthetic curled up in my palm. I tell my friend that I am going to get off the bus and stop at my uncle's house on the way to school. I get off the bus, walk the few blocks to his house, use my key to get in, and hope to make a mad dash for the bathroom. I have not counted on the chaos of a house with two teenage boys and a younger sister who enjoys tormenting them on a morning before school. These three kids, singly and in pairs, are either in or blocking the door to the bathroom.

At this point in my life I am pretty quiet and people mostly ignore me. I sit at the kitchen table, prepared to wait my turn for the bathroom. After almost a half hour, just as I am about to get into the bathroom, my teenage boy cousin squeezes in front of me and turns, laughing, to taunt me, thinking he is getting one over on me.

For one of the few times in my life, I put myself first and with force. I announce in no uncertain terms that I have been waiting twenty minutes for the bathroom, and I really need the bathroom now. My family, who hardly ever hears me speak, let alone yell, stares at me. My cousin asks, "What's wrong with you?"

At a loss to explain, I extend my hand. My new eyeball sits there in the center of my palm, staring back at everyone. In the stunned silence that follows, I calmly walk to the bathroom, shut the door, and pop my eyeball back into its socket. When I come out of the bathroom, no one has spoken a word. Perhaps no one has moved. "Bye, see you in school," I chirp, and walk out the back door.

Later that evening, my uncle stops by my grandmother's house on his way to work. He scolds me because, he claims, I nearly made his son, the one who plays football and works with power tools, faint.

This is the story no one will ever tell you about prosthetics. It's not a morality play and it teaches no useful lesson. No one will ever tell you how to discreetly pop a prosthetic back into place without making teenage boys faint. If I were writing a pamphlet on "Your New Prosthetic," I would suggest (right after the bit about the lightweight yet durable polymer) that you try to get past forgiving its imperfections. This is the one thing no one will ever suggest, but it is the one thing that will get you through this story and stories you cannot even imagine yet.

Alicia Kestrell Verlager is a recent graduate of MIT who writes about the intersections of disability and technology; you can visit her blog at http://www.livejournal.com/~kestrell.

CELL PHONES

E. Cabell Hankinson Gathman

I went to Japan in the Year of the Dragon, a terrified bleached blonde foreigner, because my heart was broken. I wanted to be alone, but it was in Tokyo in 2000 that I got my first cell phone. I don't remember how popular cell phones were in the United States at that time.

In high school, a few of my classmates had owned pagers, although we mostly associated them with drug dealing. During my college years no one in my family had a cell phone nor did my freshman or sophomore year dorm mates. As a female exchange student, however, I was housed in the Tokyo YWCA whose rooms did not have landlines. There was a single shared phone in the first floor corridor, across from the caretaker's apartment. Talking on it transported me to my imaginings of a 1950s' dormitory—it seemed a lot like prison. Receiving calls on the YMCA phone was even worse than attempting to make them, so the first thing resident students recommended to newcomers was that we get *keitai denwa*—cell phones—*keitai* for short.

I was nineteen years old, which made me a minor under Japanese law, unable to sign the *keitai* contract. Olga from Uzbekistan pretended to be my older sister and signed on my behalf. My choice of phone models was restricted by my near-total illiteracy in Japanese; I needed something bilingual, and I also wanted something pink. Fortunately I was in Japan, where even bilingual phones came in pink—and not just pink, but a softly gleaming pearlescent shade, the keys smooth and jewel-like, marked with *hiragana* and Roman alphabetic characters beside the larger numerals. It wasn't long before I'd plastered the back of it with a sparkling translucent sticker of Hello Kitty wearing a scorpion costume, denoting my Western zodiac sign. When I came back from Hiroshima, it was with the dangling addition of a souvenir *keitai* charm from Miyajima Island, a metal maple leaf in brilliant neons.

It wasn't an international cell phone really, but it was possible to purchase international calling cards for it. It let me talk to my parents and to friends. I had a laptop but no Internet at the YWCA, and it was a semester before I figured out that I could get an Ethernet cable and connect my laptop on campus instead of waiting for the horrible slow machines at the student computer lab to become available. I mostly called people very late at night while drunk, which meant that it was usually about nine o'clock in the morning in Missouri, the place I was calling. My parents were pretty much the only ones willing to pay to call me, so if I wanted to talk to my friends, I had to call them. Looking back, I can see that this might have depressed me, but at the time it suited me fine. I had run off to Japan because it was the only place far enough away that seemed safe; I took comfort in controlling with whom I was in contact.

Alone in Japan I was safe from my relational ineptitude. In Missouri, there was a guy whose neediness had smothered me and a girl who made me wonder if I would ever find someone interesting who was also nontoxic. The only time I received a call from someone outside of my family, it was a mutual friend who phoned to tell me that the needy guy and the toxic girl had moved in together. She said she thought I would "want to know." It took me three intensive Harajuku shopping trips and an all-day visit to Tokyo Disney to recover from that phone call. After that, I preferred the radio silence. It was a soothing necessity. Sometimes, for a moment, I would think I saw someone on the street I didn't want to see, but of course it was never she—always some slightly-off doppelganger, redrawn Japanese. And she didn't call.

I text messaged in Japan more than I ever have since. It was much cheaper than on American plans, and as my Japanese improved, the phonetic *kana* and pictographic *kanji* allowed me to communicate exponentially more meaning within the character limit for a single message than I ever could in English, even with the use of chat argot. Texting gave me something to do on the train, speeding through the city, never idle. In Tokyo, a world in itself, my glittering fairyland of the future, the *keitai* was an easy way to find my fellow students on the spur of the moment, colliding like lonely atoms in coffee shops and among the English language paperback racks at Kinokuniya. Throughout my travels on that colossal neon grid, the *keitai* was my talisman.

I brought it back to the United States even though it didn't work here—J-Phone, unlike its competitor Do-CoMo, had no agreement with any U.S. mobile carriers. I tucked it into a drawer with its charger, and every now and then I took it out and stroked its plastic skin, switched it on and tried to remember the world that was and the blonde girl who had moved through it, carrying this *keitai.* Tokyo sat trapped inside it: the ring tone that I'd so painstakingly programmed, the old text messages, many of them in *kanji* and *hiragana,* exchanged with the people I knew there. Unlike email, those messages could be accessed only through the single handset onto which they had been downloaded. Unlike ruby slippers, the *keitai* could never really take me back, and it could never be integrated into my American life. Like my year in Japan, it just didn't fit. I came back from Tokyo feeling in some ways as if I'd never left Missouri at all—I was different, everything was different, but somehow no one realized that the world had shifted.

Like Lucy coming out of the wardrobe in *The Chronicles of Narnia,* no one even noticed I'd been gone. My *keitai* held the voices of Japan; it was useless to try to make it talk to Americans. That interval of my life was simply past, and no one remembered it but me. If I hadn't had the phone, it might not have ever happened at all.

American cell phones were ugly and squat. I knew more people who carried them, but the landline in my dorm room was free of charge and at Truman State University in Kirksville, Missouri, where everyone I knew spent most of their time around the first-floor dorm lounge anyway, a cell seemed kind of pointless. There were no trains in Kirksville, no massive bookstores where you could lose your friends between the second floor and the fourth. I didn't need a cell phone there, and if I couldn't get a shiny pink one, then I didn't *want* one, either.

It was only two years later, when I graduated and picked up for Wisconsin, that I thought a cell phone seemed "worth it," and even then, only for practical reasons. It was the unlimited off-peak minutes that did it, the prospect of being able to make free calls on nights and weekends to people I was leaving behind. That cell phone was a very basic, navy blue Nokia, and our relationship was purely business. I carried it in my bag. It made me reachable in a strange place. I had cable Internet, and the only calls I ever got on my landline were for someone named Earl, who seemed to have run afoul of various authorities. No one believed me when I said he didn't live there anymore. I got the landline disconnected.

A couple of years after that, I switched networks and got a new Audiovox flip phone. Being able to keep my old phone number was a big deal, and the companies did everything they could to make it easy—no notifications of a changed number to be made, no relationships to be reaffirmed or discarded with the switch. I had to transfer my Nokia contact list to the new handset myself, but it seemed worth it. The new model felt like the future. It was silver, not pink, but that was still an improvement over the matte navy blue of the plastic brick I'd been carrying around. And the Audiovox had a camera. I was very, very upset when, during the first month I owned the phone, I dropped it and scratched its casing. It still worked, but I couldn't shake the feeling that a luminous future had been marred.

It was some time before I discovered the vast selection of ring tones that I could download for the Audiovox. I quickly amassed quite the collection—"Girls Just Wanna Have Fun" announced my sisters; when my academic adviser called, the phone played "Sympathy for the Devil." My default was usually the theme song from *Buffy the Vampire Slayer,* a peppy number that worked well when translated into polyphonic midi format and announced to anyone in earshot that here was a fellow *Buffy* viewer, if they cared.

I became involved with a man, an unavailable one this time—needy in his own way, but hardly willing to touch my shoulder in public. I can't really blame my cell phone, but in my heart of hearts I know it enabled our inappropriate connection. We rarely spoke in public—certainly nothing personal, no touching, nothing to betray us. He called me in the middle of the night, when we were both alone. Sometimes when I was trying to fight my way back to myself, to preserve some pride, I tried to get the cell phone to help me: the phone could be turned off, set to silent, shoved into a drawer across the apartment. Perhaps he was doing the same thing or just cared less. He didn't always answer his phone when I called—maybe half the time. But I got nervous if my cell phone was more than a yard away, afraid I might somehow miss his call, despite more than once having been woken from a relatively sound sleep by a ring tone from another room. I slept with the Audiovox at arm's reach, the volume cranked to maximum, waiting for the ring tone I'd set to herald his calls. Our song. He heard it once, when he had to call my phone to wake me up from a nap so that I would let him into my apartment—of course I was sleeping with it at my side. He commented on it later, how he heard it on the other side of the door, our song signaling his presence. Our song, rendered polyphonic, my way of having him inside my Audiovox.

Sometimes I went to the Verizon Web site just to look at the call log for my account, as if the number of minutes we'd spoken in the past week could be plugged into a formula that would yield n, where $n = $ *how much he loves me.* Once when he was angry he didn't call me for a week. I spent hours with my phone burning a lonely, desperate hole in my pajama pocket. We communicated more by email and chat than by phone, actually, but the phone calls always meant more to me; they remain relatively unspoiled in memory. He called me to whisper in my ear; he called me when I had to fly home to my dying grandmother. I saved voice mail for as long as Verizon would allow, calling back at the end of the twenty-one-day default period just to preserve them.

When he broke up with me, it wasn't by phone. He waited until I was once again in a foreign land, California, this time, away from everyone I knew, and then he broke up with me via instant messenger. We didn't speak for four months. Finally, he wanted to talk to me, but he didn't want to see me, and once again the cell phone was how he made me—how he made us—invisible. I thought at the time that he was afraid of me, that the phone was protecting him from me, but really it wasn't much different from the way the phone had always protected him—kept me secret, made me safe. Or maybe, because it was so easy to feel that I could be hidden inside his phone and inside his computer— never mind the data trail—he never had to confront whether the reality of what we had was really as bad as he thought it was. If he kept me in the invisible cracks of machines, he could have some part of me without facing me or his fears.

I took his number off my phone within hours after he'd broken it off. I had too much pride to leave myself easy access to his voice mail. I knew he'd never have the courage to pick up my calls. I didn't want to leave my own pathetic data trail. My grandmother died that same week, but I kept her number. There was still something that felt like a connection I didn't want to delete. I still have her number in my address book. His number, on the other hand, I wish I'd never known it at all. Yet I couldn't bring myself to delete his ring tone. Whenever I cycled through my store of tones, to change my default or set a tone for a particular contact, there it was, playing automatically as soon as I reached it on the list. Every time I heard it, it shocked me for a moment with instinctive pleasure, the sense that we shared something, that he wanted to talk to me, before it filtered through the present and I remembered that he hadn't even called to say he wasn't going to call anymore. I would skip past it as quickly as possible, but I always knew it was coming and didn't get rid of it. I came to think of this with a certain sense of humor: just my little tendency toward techno-mediated, psychic self-mutilation.

And then, last month, I got a new phone, an upgrade from Verizon. I consulted with my blog readership about which model to choose, first narrowing the field to two choices because of their superior built-in cameras, music-playing capabilities, and availability in pink. There were so many conflicting opinions that I finally had to go by instinct, choosing the model that seemed to have the more intuitive interface: the Motorola RAZR V3m. Its pink wasn't as vibrant as I would have liked, but there was a wide range of snap-on cases available for it, and I was able to find one in a pink and black zebra-skin pattern that seemed to be just right for me. I have also spent a lot of money on its ring tones. These are extravagant gifts to this phone and no other—unlike telephone contacts, Verizon will not let me transfer ring tones phone-to-phone. So, my dead grandmother's telephone number is in my new phone, transferred with the rest of my telephone contacts. My dead relationship's ring tone is not.

And it's not just his ring tone that's gone. He belongs to another artifact. It sometimes feels as though the year of my life I spent in that ill-advised love affair was poured into the now-obsolete Audiovox. Like a gift that still possesses something of the soul of the giver, the phone itself had come to be haunted by the voice that spoke so often through it. I gave back everything he had given me, but I couldn't get him out of my phone. I used to get tattooed when love went wrong, remaking my body into one that had never known the touch of my former partner. Now I have a new phone into which I have never spoken too soon or not wisely enough. Now I have a phone from which I have never failed to hear the words I wanted. My new RAZR is a part of a me that is freshly born, unscarred.

And I carry it everywhere. I frequently speak on it as I move through public places—I find that it provides insulation against panhandling and unwelcome advances. It's difficult, after all, to engage someone in conversation when they are already talking to somebody else. If I can't get a signal, I don't go as far as my friends who will pretend to talk to a dead phone; I just play Tetris instead. The blonde girl with her pearlescent pink *keitai* will never walk the streets of Tokyo again, but now there's a pink-haired woman with a pink zebra-skin RAZR that trills "Heartbreak Beat" for default calls in Boston.

E. Cabell Hankinson Gathman is currently a PHD student in sociology at the University of Wisconsin–Madison. Her research interests are in the use of Internet technologies—including social networking sites, virtual worlds, chat and blogs—to maintain and develop relationships.

THE PATTERNING TABLE

Nicholas A. Knouf

Our basement walls were covered with charts, schedules, and sets of instructions. In the southeast corner stood the patterning table. Made when I was eight years old by a friend of the family who lived in a nearby town, the patterning table came up to my chest. Supported on its four corners by thick wooden legs attached with massive bolts, the highly varnished table was strong enough for a fifty-to-eighty pound person to be moved around on its flat surface. That surface was padded with Naugahyde, which in turn was covered by sheepskin that fitted snugly around the edges.

At this table, every Monday through Saturday, three people (five, if we were training new volunteers) surrounded my sister, Robin. We gently took hold of her fragile arms, legs, and head. With a regular rhythm we moved her extremities in the motions of a crawl: one, turn the head to the left, bring the left arm away from the body and next to the head, bend the left leg next to the torso, and vice versa for the right side; two, keep the head where it was, pull the right arm back next to the body, extend the left leg, and vice versa for the right side; three, repeat as one, but switching left for right; four, repeat as two, but switching left for right. One, two, three, four, the count continued over the course of five minutes, and the patterning session was done, for this hour. The ding of a kitchen timer told us it was time to move on. Next hour we repeated it, on the same table with the same sheepskin cover, with different volunteers. All day this continued, according to the schedules, instructions, and charts that my mom wrote in thick strokes on butcher paper and taped to our recently sheetrocked basement walls. Patterning was combined with breathing exercises, using a mask designed to increase the blood flow to Robin's brain.

When I was in first grade Robin was diagnosed with Rett Syndrome, a rare neurological disorder that may include devastating mental and physical retardation. It afflicts only girls, most of whom are not expected to live past their eighteenth birthday. The girls often make rocking motions while sitting, wringing their hands. Many are able to develop some basic level of functioning, such as simple mobility or being able to feed themselves. Robin, however, could do none of these things; she was utterly dependent on us. Even so, there were still the smiles, tears, and frowns of any young girl, and in her eyes you could see thoughts she could not speak.

Robin's disease led our family to take on a major commitment: a full-time, in-home physical therapy program. Our location, rural Iowa, meant we had to develop everything ourselves. To build the equipment, to schedule the people who came in on a regular basis, week after week, we found volunteers, some local, some from afar.

The patterning table took up a good part of our basement, but at a certain point during Robin's therapy we set up another apparatus for her, a jungle-gym-like contraption made by the same volunteer who built the patterning table. The new contraption reached from the floor to our seven-foot ceiling. The height of the jungle-gym bars was adjustable. At certain times the overhead bars were low enough so that Robin could walk her hands along them. When she was stronger, the bars were raised and Robin had to grab onto freely swinging handholds of thick nylon rope that hung from the bars;

she wore specially designed Velcro footies that stuck to our short-pile institutional carpeting. The resistance from the Velcro was meant to make it harder for her to lift her foot, while the swinging rope was meant to improve her sense of balance. It was hoped that this would strengthen Robin's legs and upper body in preparation for walking on her own.

Our work at the patterning table was based on the Doman-Delacato method. It laid out a series of therapies designed to help brain-damaged children toward better functioning. The method is based on an evolutionary metaphor: the individual develops in their lifetime just as the human evolved over generations. That is, development begins at the fish and reptilian stage (crawling) and moves through to mammals and primates (creeping, with the stomach off the ground). Doman-Delacato reasons that if you "pattern" a brain-damaged child with the motions that are involved in each of these stages, you will unlock the later stages of development. So the repetitive motions on the patterning table were supposed to teach Robin's muscle memory to crawl. Yet she never was able to crawl on her own: she learned only to walk a step or two, lightly aided.

The Doman-Delacato method has not found empirical support, and the American Academy of Pediatrics has issued a number of warnings about the technique over the years. As a child I never questioned our program. To me, it seemed natural that if we moved Robin's arms and legs in a certain manner, we eventually would "train" her brain to move body parts on its own. And of course my parents, like so many others, seized on the program's marketing and the bits of anecdotal evidence that suggested it worked. The Doman-Delacato method provided hope, but I think back on it now with sadness, regret, anger, and resignation.

Robin began to experience seizures when I was in fifth grade, not unusual for Rett girls who often regress from their developmental plateau. I remember the huge stuffed rabbit that lay on the table the day Robin was rushed to the hospital because of her first, unexpected seizure. The seizures made it impossible to continue with the treatments on the patterning table. Yet the patterning table remained in the basement. We removed the sheepskin, and then the Naugahyde quilting, and used the table as any other flat surface, a place to store papers, books, and mail. Gradually the schedules and charts on the butcher-block paper began to come off our walls, but the patterning table remained, a reminder of what we had done, of what we had tried to do.

Even as a young child, I read the few journal articles our family had about Rett Syndrome. They were over my head, but I persisted in my efforts to figure out what was going on with my sister. I launched into my parents' popular science book on genetics. A year or two before I took any proper biology class, I was discovering genotypes, phenotypes, and karyotypes, the genetic bases of diseases such as sickle-cell anemia and Parkinson's. At the time, the Human Genome Project was in its nascent stages and the hope for "miracle" cures was strong. With the insufficient knowledge I had, and bolstered by the arguments in the genetics book, I believed that finding the gene or genes that "caused" her disease would lead immediately to gene therapy that would "reverse" Robin's genetic malfunction. Of course you had to have a way to get the genes into the cells, so I eagerly read about techniques to force the existing chromatin to take up new genetic material—retroviruses and exotic "electro gene therapy." Then came the problem of how to reverse the damage in existing cells. Neuron regrowth in the brain is meager at best; there, repairing genes would not be enough.

As a middle school student growing up with my sister, I thought these problems surmountable. And I determined that all of this would be my doctoral work:

I would find the gene that caused Rett Syndrome and discover the necessary therapies to cure the disease. It was merely a matter of time. The summer before high school, I enrolled in a summer program in molecular biology at the state university, a first chance to work with the tools I had read about: restriction enzymes, plasmids, and ethidium bromide stains. I was in the throes of passion: I pored over books I could not understand; I persuaded other students to work unattended in laboratories full of dangerous reagents. I was in a hurry.

But Robin died in the fall of 1993, just short of her ninth birthday. I sunk into a world of grey and black, my schoolwork the only thing that kept me going. I remember the rush to finish assignments in my high-school biology course that were late because of Robin's funeral and the time I took off after her death. I remember the fleeting thought that if I could complete my work faster, I could start college early and be on my way to finding the cure for other Rett girls.

I went to Caltech as an undergraduate to study molecular biology with one sole objective in mind: to find a cure for Robin's disease. In high school I had performed in a chamber music group and had notions about attending a conservatory. I devoured literature and thought about being an English major and studying philosophy. But now, stronger than all of these were my memories of Robin and the patterning table. I was driven to discover the gene for Rett syndrome. I saw it as an achievable goal.

Then, on a gray California winter day during my sophomore year, I read that a group of researchers had discovered the gene responsible for Rett Syndrome. I posted the article outside my room. I told all of my friends. I wanted to be among the people in the article. But there was more work for me: turning the genetic discovery into a treatment for Rett patients. Soon after, experience as a research assistant convinced me that I was not suited for the biology workbench. I moved toward cognitive sciences, with its higher-level descriptions of perception, action, and emotion.

The patterning table, long unused even during Robin's life, was finally removed from our basement, probably passed on to another family using the Doman-Delacato method. Its absence left a space. I had passed the patterning table and its volunteers in the early morning; I had looked toward it when I came home from school; I had walked to it when it was my turn to be part of the group that moved Robin's head, legs, and arms. Without the steady presence of the table, where would we turn? Where would our efforts be channeled? Without Robin's influence, what would be our purpose? The table was more than a focus for our thoughts; it anchored our love for Robin and the energy we put into giving her a future. Like the tables in traditional Midwestern churches or cafes, the volunteers who assembled chatted and gossiped and spoke about their lives. It brought Robin into a community. When Robin was happy and obedient, her attendants gave her praise; when she was cranky and ornery, they understood, but lightly scolded. The patterning table made life coherent, all of a piece. With Robin, the volunteers, the discredited therapy method, and the patterning table, we had tried to awaken cognition with care, with the soft sheepskin, the men and women gathered around.

Nicholas A. Knouf is a doctoral student at Cornell University working on the interdisciplinary study of robotics, critical theory, sound, and science, technology, and society. He received his master's degree in Media Arts and Sciences from the Media Lab at MIT.

TELEVISION

Orit Kuritsky-Fox

The first time I watched television I was four; my mother and I went to visit the downstairs neighbors. On the screen of their newly acquired TV set, people were stepping into mirrors and reemerging on the other sides, into new rooms with different mirrors.

They tarried there long enough to explore, and then crossed back to their starting points. I held my breath every time these people were at a threshold, imagining my own skin submerged in the mirror's thick grey mercury-like substance. When the people safely passed to the other side, I inhaled deeply, as if emerging from a dive in deep water. In the years that followed, I would do that every time I watched one of this program's mirror-crossing segments, a program I later learned was called *Vision On* and produced by the BBC.[1] I have no memory of what the neighbor's television looked like. I can recall it only as a blurred shape, reduced to another mirror frame whose purpose was to be passed through.

But I do remember a twist in that first episode: one of the characters—a tall, bearded man—walked into an elongated mirror in his living room, but to my horror, when he crossed back, he found himself in a different room. He continued through another mirror into another room, and then another and another. The chain of mirror crossings ended outside, on a manicured lawn. The scene now dissolved to a slapstick sequence: someone (or something) was chasing the bearded man, and eventually he plunged, headfirst, into a pond that reflected his image. He then emerged, wet, from the mirror—back in his living room.

Parents and Children

Elevated on a cart, the television in my grandparents' living room had an entire wall to itself. In front of it were one cushy armchair (for my grandfather) and a smaller, less comfortable one (for my grandmother). To the side was an armless sofa, on which my grandfather would lie down after work, watching television, always in the dark, always at full volume. Decorated with doilies and crystal swans, my grandparents' successive television sets were always the biggest and the most advanced models available; as soon as color broadcasting was introduced, they bought a color TV. This commitment to cutting-edge technology was quite out of place in this house, with its plain old furniture and paint chipping from the walls. It was a house where nothing was wasted; the use of food, gas, electricity, and even toilet paper was scrutinized and carefully measured. The television was exempt from this Spartan regime. Baudrillard would say that it certified "citizenship"; it was "a token of recognition, of integration, of social legitimacy."[2] My grandparents were shipped, together with other WWII displaced persons, on a cargo vessel from the port of Marseilles to the newly established state of Israel, spent several months in a tent city, and then, one day, without notice, were sent to a small town close to the northern border. They did not have money or position. For them, Israeli television was the only way to take part in any kind of public discourse.

In my own parents' living room, the TV was never a focal point. Our television set was always smaller than those of our neighbors, lost in a constellation of books, CDs, and records, as well as some "ethnic" or "antique" pieces collected by my mother in Mexico City, Venice, Paris, and Delhi. My brothers and I made fun of this room—looming large on the left was our enormous upright piano (which had pride of place long after we all stopped playing it). To its right were two parallel Corbusier-style sofas, sandwiching a sleek curved-glass coffee table, and then the shelves on which the TV was half-buried. This meant that we could not sit on a sofa and watch the TV without twisting our necks. To watch comfortably, we had to slide into a horizontal position.

We used the living room mainly for watching television, but the television itself was not allowed to be at its center. The room reflected a cultural cosmology that put high culture far above popular culture. The furniture arrangement was pedagogic. It made it clear that we were not, had never been, and never would be, riffraff.

Neither of my two uncles, members of the very religious Slabodka community, had a living room. They lived in small homes in Bnei-Brak, the capital of Israeli ultra-orthodoxy; they had many children and no extra space. And neither of them ever owned a television. Jewish ultra-orthodoxy follows a literal interpretation of the fourth commandment that bans any "graven image" or "likeness of any thing that is in heaven above, or that is in the earth beneath, or that is in the water under the earth." Television, purveyor of images, is considered an abomination in its very essence; additionally, its content is "repugnant." In religious neighborhoods, admonitory posters (called *pashkevils*) warn of its corruption.[3]

Beyond the strictures of the fourth commandment, television is absent from my uncles' homes for the same reason my grandparents made it the center of their lives: it stands for cultural participation. The Slabodka community, like other ultra-orthodoxies, defines itself through strict withdrawal from the Israeli secular mainstream. TV programs offer glimpses of other ways of life and reinforce the idea of citizenship, rejected by this community with its pre-Enlightenment worldview.

Scriptwriting

I was still officially in college when I wrote my first TV script. Through friends of friends, I was told that an editor of a show on the Children's Channel was looking for new writers. The show had a shoestring budget: one host at a time, a blue screen, and a video game that child viewers could play from home, using a telephone keypad. I watched the show a few times and decided to give scriptwriting a chance. My second script was chosen for broadcast, and I watched it air from the studio control room, a place of seeming chaos but from which, miraculously, a spectacle emerged that followed the exact order designated by my script.

In the following years I came to love this hard-to-parse cacophony. I soon became an editor and reveled in the adrenaline rush of live broadcast. While the show was on the air, my job was to talk to the hosts through small microphones hidden in their ears, giving instructions that would help them interview guests, interact with viewers, and tackle unexpected hurdles. Being in the control room felt like being part of the operation of a magic box. I was a tiny gear in a hidden mechanism buried in the belly of a shiny toy. I was a homunculus playing blindfold chess inside an automaton. Each script caused the studio beneath the control room to take different forms—a detective agency, an arcade—and to exist at different times. One show was set during the French Revolution, another during the burning of Moscow by Napoleon.

One day, getting back from the studio after a show, I found a message from my father on my answering machine. My maternal grandmother had died. Her funeral would take place the next day in a special cemetery for ultraorthodox people, located in the industrial outskirts of a Tel Aviv suburb. I knew that a drama was taking place around the funeral arrangements, a drama in production for the many years since my mother and her two younger brothers, like the characters on the show *Vision On,* passed through different thresholds, but in this new case, never returned to their common starting point.

The first thing on my mind was what to wear to the funeral. This had nothing to do with fashion; every encounter with my ultraorthodox family required careful planning and acts of impersonation. This gathering had the extra weight of being public; I assumed that some members of the Slabodka community would attend the funeral, and I didn't want to be the cause of embarrassment. I decided to go with the longest dress I had in my closet, a black sleeveless dress with a keyhole décolletage that had been to a wild party or two. I covered it with a black T-shirt to mask its boldness, and topped everything with a long-sleeved grey blouse. Feeling that wearing all of these in the middle of July was quite a sacrifice, I exempted myself from the mandatory pantyhose that my female cousins had worn year round, rain or shine, since their early childhood. I told myself that the dress is long enough; no one will notice. Since I wasn't married, I was excused from covering my head with a wig (required of married ultraorthodox women). When I looked in the mirror I noticed that my posture was transformed: my shoulders rose up, and my upper back leaned a little forward.

I arrived at the cemetery early. The funeral procession (with my parents and brothers) was delayed in my uncles' town, whose residents had prepared pre-funeral eulogies. Unknown to most Israelis, this cemetery, like an exclusive country club, is a friend-brings-a-friend place. It guarantees fastidious compliance with Jewish burial laws. The experience begins with the separation of men and women in its reception area.

My grandmother would not herself have cared about this meticulous orthodoxy. She followed some basic religious rules but only to appease her two sons and husband (who didn't follow a very orthodox way of life himself but didn't want to stand out in their community). She was never initiated into orthodoxy; as a young orphan growing up in Warsaw she had to work as a seamstress to support her siblings and the woman who had adopted them. Under these circumstances, kosher food and observing the Shabbat were unaffordable luxuries. As soon as the Second World War started, she was constantly on the run and ended up in a Russian forced labor camp in Finland, together with my grandfather. After my grandparents settled in Israel, they aligned themselves with the Zionist religious movement, which offered a middle ground between their Jewish traditional past (both real and as a nostalgic construct) and the Zionist establishment whose commitments were political, not religious—an end to the Jewish Diaspora. But for the ultraorthodox sects that use the cemetery she was to be buried in, the Zionist religious movement to which my grandmother belonged is an adversary. The ultraorthodox see it as watering down Judaism and manipulating divine prerogative. For them, Zionism transgresses the prohibition that one must not hasten the day of redemption.

As I was waiting at the cemetery, the receptionist approached me and asked what I was looking for. I told him that I was there for my grandmother's funeral. He stared at me with dismay. I thought to myself that perhaps the pantyhose had not been dispensable. "Are you religious?" he asked, and I started to mumble. Then, he found a way to cut to the chase: "Do you own a television?" I hesitated for a second and then told him the truth. He gasped, looked into my eyes and asked his next question: "Your grandmother, she didn't watch television, did she?"

I knew I'd save everybody a lot of trouble if I lied. I was about to, but then I remembered my mother. Our relationship at the time was tempestuous, but this white lie felt like sheer betrayal. I told the old gentlemen the truth.

My mother arrived at the cemetery with my father and brothers. She was dressed in huge dark sunglasses, a colorful veil, and a very short pencil skirt. The funeral party was large: seven buses opened their doors and masses of yeshiva students dressed in black suits got out of them. The cemetery personnel looked confused; there was me (sans pantyhose) and my mother (a bohemian Jackie O), yet there was the massive traffic of orthodox mourners, usually associated with the funerals of high-ranking prodigy rabbis.

The whispering started immediately. The receptionist took my uncles aside and spoke with them while occasionally pointing at me. Then he pointed to my mother, who was now attracting everyone's gaze, like a human lightning rod. They talked for a while, nodding their heads, and parted ways. My uncles' wives rushed to my brothers and me to say that if asked, we must insist that our grandparents actually never owned a television, that all of this was a big misunderstanding. Then came the eulogies praising my grandmother's religious virtues, and I couldn't help but smile and think of my grandparents' living room and of my grandmother, transported, passing through thresholds, speaking Yiddish with the news anchors, and offering family advice to the characters on *Dynasty* and *The Bold and the Beautiful.*

Orit Kuritsky-Fox is a graduate student in MIT's Comparative Media Studies Program. She has been a TV scriptwriter and editor in her native Israel, and a producer for the syndicated talk show "The Connection" that aired from the NPR affiliate WBUR, Boston.

Through Clinical Practice

THE WORLD WIDE WEB

John Hamilton

As the senior child psychiatrist at a busy clinical center in northern California, I interview troubled male adolescents daily. Many, usually referred by other clinicians, have suffered serious depressions and made suicide attempts. Yet despite the seriousness of these cases, when I speak to these boys' parents, they often bring up another complaint: "I can't get him off the computer."

In my office, these young men are withdrawn and hard to reach. But many have active virtual lives in which they are energetic and forthcoming. These virtual lives represent an opportunity for the patient and for the clinician. Clinicians need to understand what these young men are doing online and find ways to use this material in the therapeutic encounter. Yet all too often, clinicians dismiss online activity with an attitude characterized by the phrases "don't let them on the Internet" or "tell them to get a real friend." This approach is worse than unproductive because it denies the importance of what may be the liveliest part of patients' realities.

On a practical level, adolescents spend so much time on the World Wide Web—writing blogs, hanging out in chat rooms, participating in social networking sites—that the clinician will often not understand what patients are talking about without knowledge of their online worlds. On a deeper level, bringing the

World Wide Web into the therapy session offers new opportunities for understanding patients' inner lives as well as deepening and exploring the therapist-patient relationship.[1] In particular, online life offers a window onto the transference, the feelings that the patient brings to the therapist from other relationships. Today's young male patients often have impaired, absent, hostile, or deceased fathers. It is not surprising that these patients should develop negative transferences toward their therapists, transferences that need to be clarified and verbalized for the therapy to survive and progress. This is made more difficult if therapists actively deny a crucial aspect of these boys' reality.

In my practice I find that bringing the Internet into the therapy session enables difficult things to be said. The Internet takes a therapist and patient struggling to communicate only with words and offers them color, sound, and mobile avatars. The endless variety of Internet sites makes it possible for young men to find particular places and games that help them work on their inner life. They are even able to find characters to "play" that help them address specific psychological issues.

Here I focus on how early adolescent boys use the Internet to help them develop feelings of appropriate masculinity. Some use therapy to discuss their relationships to online heroic figures.[2] On the Internet, many of my patients play the parts of timeless masculine heroes, men who act alone, like to fight, have powerful, sculpted bodies, take extreme risks, and are allowed sexual adventure only if it does not lead to emotional intimacy. The heroes avoid self-reflection, associated with lethal doubt.[3] The choice of these heroic fantasy figures, virtual hypermales, comes with a cost. The heroes embody exaggerated masculine traits. Successful adolescents, living in reality and not the virtual world, do better when they collaborate, share experiences, understand others, and see nuance—all skills that hypermale super-heroes lack.[4] In my therapy with patients who identify with virtual hypermales, we challenge the simple equation of masculinity with the hypermales' aggression and lack of communication.

Role Playing

Thirteen-year-old Eric is in crisis when he comes to see me. He looks older than his age and has an unchanging, anxious smile. He has already been diagnosed with Attention Deficit Hyperactivity Disorder (ADHD), as well as Asperger's Syndrome, generally thought to be a high-functioning form of autism that makes it hard to be socially attuned to other people. Academically, Eric has done very well in elementary school and tested with a full scale IQ of 140, but now, in junior high school, he is failing.

When I meet Eric and hear his story, I recognize many factors that could have contributed to his problems. He went through an early surgery to remove an atrophied testis, a difficult intervention for a young man to deal with. His father is an openly hostile man who works at night and sleeps during the day and never happily engages Eric in shared activities. Eric's interactions with his father are emotionally distant. Drawn to the Internet, Eric faces peers who bully him and parents who dismiss his preoccupation with fantasy characters and their exploits.

Eric says he feels "hurt" that he is unable to make friends. He has become a confirmed loner; when he thinks of others, his thoughts turn toward violence. He told his seventh-grade teacher, "If a person is depressed, he shouldn't kill himself but should kill the person who is giving him trouble," a comment that first got Eric suspended and then put into a special class. Eric's favorite hobby is cartooning, a hobby that his father openly mocks. In fact, Eric's cartooning is impressive. He has created and drawn all the art for a role-playing game with twenty interacting characters.

Eric carries a portfolio of his cartoon work to our first therapy session. In response to my questions, he describes each of his hand-drawn characters. Most are eroticized hybrids of males, females, and animals. One female character, Allison, has a tail like a dog; another character, Alex, can be either a girl or a boy; Will and Reni were made in a lab, separated at birth, and reunited at age fourteen; other characters are gay, bisexual, or actively homophobic. Eric says his drawings are often bisexual because he likes his characters to have sex, and he sees bisexuality as leading to more sexual activity.

By the end of our first session, Eric and I establish what becomes our routine: after a minute or two of conversation, Eric goes over to my office computer and shows me the sites he visits when he is bored. Although he seems anxious about bringing me into his private world, he does so and in great detail. The sites Eric shows me have much in common. Each is built around mocking or morbid humor and includes comics, games, and movie clips. One plays with the themes and characters of the film *Lord of the Rings*.[5] On this site, Eric shows me one of his favorite characters, Faldon, who Eric describes as "keep[ing] his robe so white by retreating quickly" and by "having no superpowers." As Sherry Turkle has noted, what people do with their time on the Internet can have the qualities of a Rorschach, serving as a projective screen for inner preoccupations. And indeed, understanding Faldon is a way to get to know Eric.[6] In his preoccupation with the seriously flawed Faldon, I can see Eric expressing his wish for the masculine heroic and his sense that both he and his father fall short.

In my office, Eric frequently visits the site for *Bob and George* comics. Its hero, Megaman, is something of an anti-hero, mocked by his creator, who draws himself into the strip. Eric tries to explain the storyline to me. Megaman has a brother, Protoman, and both are on the team of Dr. Light, a good guy. Megaman's past is represented by an 8-bit (and therefore crudely drawn) figure; Megaman's present is represented by a 16-bit figure; and a 32-bit (and therefore very well defined) figure shows his future. Both Megaman and Protoman have human faces and can talk. When they are hit, they feel pain and bleed. Although part human, each is also part robot. Each can change a hand into a pistol. Each is the master of smaller, weaker robots. Dr. Light's team has Megaman and Protoman, the good guys; Dr. Wiley's team contains Quickman, Turboman, Heatman, and Fireman—these are the bad guys.

Eric is deeply involved with Megaman's story. He enjoys its backstory for its intricacy and cast of dark and light characters. For Eric, it is appealing that Megaman characters are part robot. It makes them more stoic, and therefore more masculine.[7] I ask Eric to choose which character he would like to be and he chooses Shinyman, who has emeralds in his head and rubies in his hair. Eric thinks that Shinyman will someday get blown up, a common fate for action characters. Eric and I discuss how Shinyman may find himself popular because many people like emeralds and rubies.

Like Megaman, Shinyman's appeal is that as the game proceeds, he can morph into increasingly powerful forms, represented by more sharply defined graphical figures, changes that I see as mirroring the evolving self of the adolescent. Megaman and Shinyman may be simple characters, but they help Eric by providing a stable baseline from which he can construct a more complex masculine self. The place that provides stability becomes a place to experiment with greater possibility. As we work together, Eric is able to show me how the characters on his favorite Web sites have dimension, complexity, and history—some even exist in parallel universes on the Net. I am able to use this material to suggest that having an alternate universe is useful to game characters and that Eric's having the Internet as a parallel universe could be useful as well. As treatment progresses, Eric problem-solves for avatars in the virtual and takes what he learns back into the real.

After a few of our therapy sessions, there is a crisis. Eric's mother calls to say that he has written a violent story in English class. The story takes place in a school meeting, during which a boy stabs everyone in the class except an imaginary man with a clock. After he wrote the story, Eric asked his English teacher, "Are you afraid of me?" Our sessions gave him permission to express his feelings; he took that permission to another venue—creative writing. For Eric, the line between our sessions and the rest of life had become blurred. My status, too, may have moved from real to imaginary.[8] I have a prominent clock in my office, and Eric often doesn't want to end our sessions; it makes sense that he would spare me in a violent scene. I see it as a good sign that if I am the imaginary man, I have witnessed but escaped the fantasy stabbing.

In response to this incident, Eric's school administration puts him into a special education class and advises Eric's parents that he should be kept off the Internet at home. His parents comply; they have little choice, but agree that Eric and I may use the Internet together in our sessions. With this new constraint on his behavior at home, our sessions take on new urgency: when we next meet, Eric describes his latest cartoons in a hurry, eager to go online as soon as possible.

My goal is to help Eric recognize and verbalize emotional states in himself and others by identifying them first in Web-based figures. Once I have a basic grasp of Megaman's world, I am able to help Eric imagine the possible emotional states of game characters. I work with Internet characters much the way a child therapist works with toy dolls. Eric explains the characters to me and seems to enjoy the status of being the doctor's teacher. He describes how Megaman can merge with Nanobots to become Megahulk, how Karzak, the Lord of Hatred and Lord of Pain, hates everything, and how the tortured Cutman repeatedly stabs himself and others. I think that Cutman may well have been the model for Eric's story about stabbing other people.

I focus Eric's attention on Karzak and Cutman, since each is defined by a difficult emotional state (Karzak by hate, Cutman by rage and depression). I ask Eric whom Karzak hates. Eric responds, "Everybody." I ask how Karzak became so bitter. Eric responds, "People dumped on him." "Yeah," I say, "it makes sense to me he'd be really mad and even feel hate a lot of the time." The therapy advances through our discussions of the Megaman site. Eric and I now have Karzak and Cutman as symbolic markers for hate and anger and depression. For the first time, we have a common language for emotional states that could result from bullying and social isolation. I work on a common language for sexual matters as well. When Eric says of his drawings, "You can't tell if they are a man or a woman," I respond that you can tell a man from a woman because a man has a penis and a woman a vagina. Eric replies that he is amazed I would say that. My comment also identifies Eric as masculine since his penis is intact despite the fact that he has had one testicle removed.

At this point in Eric's treatment, his mother reports that Eric is showing some interest in two neighborhood girls. He wants to visit them. His mother tells me that this is Eric's first stated interest in children his own age. Eric's younger sister, eleven, teases him about having a girlfriend.

Now Eric's smile no longer seems as fixed. He brings in many new drawings and a book about fantasy chronicles of the middle ages. His parents restore his online privileges, and Eric shows me a story he has posted to the Internet. Previously, Eric accessed other people's online drawing and writing. Now he is sharing his own. Eric's online boldness has a corollary in his way of approaching his family. In our sessions he is more able to move beyond fantasy to discuss how, in reality, he is often unhappy with them. For example, he mentions that he doesn't like going to Home Depot, a frequent family destination. In a meeting at which his mother is present, the three of us conclude that Eric would be more cooperative in family matters if he were given more of a voice in choosing destinations for family excursions. At the end of this session, I encourage Eric to "have a voice about things" that matter to him.

While waiting for a subsequent session to begin, Eric's mother picks up a book in my waiting room and reads to her son as though he were a much younger child. Eric tolerates the reading, but begins his time with me by saying with a smile, "You're saving me from my mother! She's reading to me." This is the most direct and relevant comment Eric has ever made about his family relationships. At the same session, Eric sits beside me at my desk and shows me his stories as he downloads them from the Web to my computer. Now, Eric writes many stories. He adopts the ironic, self-mocking tone of the Bob and George comics and adds an intensely erotic story line. He takes ideas from *The Lord of the Rings, Harry Potter,* and the role-playing board game Dungeons and Dragons and gives them his own spin.

One of Eric's stories involves two women, Alpha, defined by "curves and her E-cup bust," and Omega, scantily clad but "quite flat-chested." Both sing siren songs, but "every psycho heard Alpha's song while every cool guy heard Omega's." Two men answer their songs. One swears a lot, a "bad ass with a motorcycle that can transform into a killer robot"; the other has "main character power!" Together these two figures form a Fellowship. I find this promising. I think that Eric is becoming grounded by writing and telling the stories of his own life and he is pairing exaggerated masculinity with "main character power." If Freud were alive, he might say that the motorcycle id and the main character ego are finally talking to each other. Offline, things are also going well. Eric is talking with his classmates and is beginning to do well academically.

Eric and I discuss his mother's disapproval of the most explicitly erotic parts of his stories. I tell him it is a "guy thing" to like "hot" stories, and we strategize about how Eric can deal with his mother's standards. I suggest he could have a "mom-approved version" of his stories and still use whatever language he himself chooses for our meetings. Eric seems to welcome this idea.

In the course of our work, Eric explores masculinity online and becomes more competent in verbalizing his feelings about hate, rage, and sexuality, sometimes by playing characters, sometimes through stories. In the course of all this, we achieve a new level of intimacy. I work with Eric to articulate what is both appealing and problematic about hypermale characters. Which aspects of hypermale characters does he wish to emulate? Which might get him in trouble? Which might inhibit relationship? I use online avatars to provide a shorthand for feelings and for opening conversations about them. I do not look at avatars as an escape; I see them as a new tool for developing self-reflective skills.

A younger patient, Roland, age eleven, uses avatars in a similar way. Roland is depressed but unresponsive to antidepressants. Through avatars, he is able to communicate feelings; during our sessions we try to use avatars to understand the feelings of others. I am growing frustrated with the case because it is clear that Roland's father is having angry blowups at home, but instead of dealing with his own rage, wants to focus the treatment on Roland. At home, Roland is withdrawn; during our sessions on the Internet, we go to the site of the *Dragon Ball Z,* where Supersonic warriors are fierce and aggressive. During one session, I ask Roland about the warriors: Who is angry and who is frustrated? Who is angriest? What is happening when they stare aggressively at each other, eyeball to eyeball? He is able to talk about their hatreds. During the following session his mother tells me that Roland is doing much better and is less grumpy. Another patient, Tomasio, age fifteen, is afraid of his attraction to girls at his school and afraid even to talk about it. He shows me a site that is a virtual high school. There, his character "tames" girls by having sex with them.[9] We use a discussion of this avatar to talk about his own sexual anxieties.

Here I am arguing for therapists' active engagement in patients' online lives. Working with Web sites and avatars is only one path. For example, my work with Manu, a seventeen-year-old Asian-American boy, centers on the experience of blogging.

Blogging

I meet Manu only days after he tried to commit suicide with a drug overdose. When I first see him, he is depressed and deeply withdrawn. I am concerned about his desire to die and worry that the hospital has made a mistake in releasing him.

In our first interview, Manu barely says a word, but he begins to speak when I ask him if he has any favorite Internet sites. Manu mentions a few music sites and tells me that he has his own Web page. After asking and receiving permission to bring up his site, we look at it together, an efficient way to get to know him. His page has photographs and drawings of his teachers, school friends, and family. It has links to his favorite music (mostly rappers), as well as a blog with almost daily entries, most written as rap verses. The photographs on his site introduce me to his teachers, neighbors, and the rich ethnic culture of his peers. I learn their nicknames; I see them in comical poses.

Beyond showing me his social world, Manu's Web page is an introduction to his inner life, including the tensions between himself and his father. Manu's father had been a lawyer in Asia who, with great hopes, immigrated to the United States only to face professional licensing difficulties. Embittered, Manu's father now drives a four-hour round trip to work as a legal associate. On his Web page, Manu expresses love for his father as well as the distance between them ("been more than a month since I've talked to him"). In one of his sketches, Manu has drawn himself wearing an NBA hat, a global common denominator of masculine identity. The background for the portrait is a series of Greco-Roman pillars that suggest a temple. Beneath Manu's drawing of his face it says, "Praise Manuism." He humorously defines "Manuism" as "the religion of worshipping me."

I am moved by Manu's Web page. It is a window onto who he was before his depression: a lively, socially engaged youth with ongoing interests in rap music, poetry, and what his site calls "peace in the middle yeast." Sharing Manu's online life also demedicalizes our encounter. We have left the institutional consulting room far behind. By clicking on a song, by pointing me to lyrics he has written only a few days before, Manu lets me get close to him. In this one session, his site provides me with a vision of how well he could do if he recovers. Beyond this, Manu's blog suggests that his distant father might be a focus for treatment, something I suggest to the leader of the group therapy program at the hospital. Simultaneously Manu begins to take medication, a course of fluoxetine. Within a month, he improves dramatically and puts on a live performance of his music for his therapy group.

According to psychoanalytic self-psychologist Heinz Kohut, an early adolescent boy begins to form a masculine identity by identifying with a healthy father who takes the role of a "self-object."[10] Deficits in fathering—sometimes due to a father's death, disinterest or physical absence—can lead to the boy's withdrawal and social isolation. Without an adequate self-object, the boy is blocked in his development. He may turn to the Internet and fantasy for representations of the masculine and get stuck there. This was the case for Eric and Manu. Treatment involves helping these boys use the Internet to communicate feelings and ideas that are hard to talk about by offering them a relationship with a responsive self-object (in this case, the therapist). In the presence of the therapist, they can work on their masculine identities. On the Internet, these boys tend to revert to what one might view as a young child's view of his father: an exaggerated masculinity of total dominance and fantastic powers. The therapist can help patients move from this stereotypical, narrow view to a more subtle and complex one. The Internet is a laboratory for male teens; when the male therapist joins the patient online it can become their shared masculine adventure. An Internet-mediated phase of treatment can serve as prologue to a more traditional "talk therapy" that mobilizes empathy, discussions of feelings toward the therapist, and efforts at joint problem solving.

Being the Main Character

A father is the first hero for his son and, in the best of times, even the mundane adventures of a dad (his work, his fixing things around the house, his way of making you and your family feel safe) are able to compete with the cinematic adventures of action heroes, far more exciting but far more distant. In my consulting room, however, I am meeting families in which the father is in no position to compete. These fathers have lost jobs and their work takes them away from their families. They are cast down by the difficulties of their lives. Their sons pay a price: in their hunger for an experience of maleness, they turn to online models whose masculinity emphasizes solitude as well as violent speech and action.

Psychotherapy is in a position to help young men understand the appeal of hypermales and to enjoy this identification without resorting to violence and without giving up constructive connections with other people. Recall how Eric gradually did become his own "main character." Manu also found his voice, in his case by revisiting with me the rich material he had been able to put on his Web site before being overwhelmed by depression. As a therapist I was able to help them realize that, while it is easy to recognize the heroism of a virtual character, it is harder but more rewarding to recognize the heroic dimensions of a human life.

These days, it is my standard practice to include Web-based material both in evaluating and treating adolescent boys. When I do so, boys suggest viewing their favorite sites, choose predefined avatars to play, create their own avatars, post writings, or even build entire Web sites. For many withdrawn boys with physically or emotionally absent fathers, real life does not offer comparable opportunities for active experimentation with masculinity. And so, it is not surprising that for these young men the virtual world is irresistible. Therapists who join them there will be amply rewarded.

John Hamilton, MD, MSC, is a child and adolescent psychiatrist with a special interest in how adolescents play and develop in both the real and virtual worlds. He is a Senior Physician with the Permanente Medical Group, Inc., in northern California.

COMPUTER GAMES

Marsha H. Levy-Warren

Tamara, a withdrawn young woman of sixteen, sits in my office. I find myself asking one question after another. Her responses are most often monosyllabic. I sense her frustration and my own.

In one exchange, about her life-long struggle with sleep, I ask her what she does when she can't sleep. She says that she spends the night on the Internet "talking."

"Who do you communicate with on the Net?"

"Lots of people."

So this silent young woman is talkative on the Internet. I am curious about this other self. I ask her if she would like my email address.

"Yeah, sure."

What evolves is remarkable in my thirty-year clinical experience: the quiet Tamara I know in the office begins to write to me almost every night. In this correspondence, she introduces me to another version of herself. Taciturn in person, in her emails Tamara is funny, chatty, insightful, charming. Her messages are usually brief, but their contents are far-ranging. She writes about her parents and her friends and her struggles with teachers. Yet during all this, she remains silent in the office.

For weeks, my efforts to talk in person with Tamara about her emails prove futile. When I mention something from an email, she tells me she doesn't want to talk about it. A turning point comes late one night, online. I tell Tamara I am tired and need to go to sleep, but that I want to continue our exchange about an incident she has just described in an email. I ask if we might follow up in my office. To my surprise, she responds with an email that says: "Sure, why not?"

Slowly, a bridge forms between who Tamara is in her emails and who she is in my office. I ask Tamara if she can see the difference between the Internet version of herself and the shy girl she initially presented to me. She nods, but points out that she has brought the more assertive self into our face-to-face relationship: "I'm here now, like this, right?"

Writing in the psychodynamic tradition, Erik Erikson describes how adolescents concretely imagine a future, stressing the importance of visualization in this playful process.[1] Using the Internet, Tamara both imagines and plays at a future, more expressive self. In our therapy, she practices it. The therapeutic exchange is a place where the imaginary becomes real.

In adolescence a stable sense of self is disrupted: adolescents change both inside and out, and need to reassess who they are once these changes have begun.[2] These days, adolescents use the Internet and computer gaming in this work, and psychotherapy is enriched by tapping into their significant use of media. Clinicians may go online with adolescents in the consulting room or, at the very least, talk with them about their online lives.

Psychoanalyst D. W. Winnicott describes his therapeutic practice as a creative play space, a transitional space in which children's development can occur.[3] This is the kind of space today's therapy can achieve with active exploration of an adolescent's use of the Internet and computer games. Online choices and experiences can be examined, elaborated, and worked through.

Computer games occupy a unique position in adolescents' lives, a space between reality and imagination. In that space, adolescents play at being who they are becoming; in therapy the elements of that play can be articulated and refined. With the increasing maturity of adolescence, young people understand more about the complexity of the outside world and who they can become in it.[4] Adolescents use computational play to make sense of the world at the same time that they escape from it. In creating ideal selves in games, they imagine themselves as if they possessed characteristics that might embolden them to approach those to whom they are attracted, or fight those by whom they feel aggrieved.

Erikson describes two theories of play: trauma and functional theories.[5] Trauma theories describe play as a place to release emotions that were repressed when individuals were overwhelmed in the past. The individual repeats the traumatic situation in play to master what was originally overwhelming or to create a new outcome.[6] In contrast, functional theories focus on play as a means for an individual to apply new capacities, for example, intellectual capacities, to the experience of their trauma. What was once beyond understanding or acceptance becomes comprehensible and bearable.

Tamara's life on the Internet exemplifies both kinds of play. In fantasy gaming, she relives prior trauma with new mastery and new intellectual and emotional capacities. In one game, she plays a warrior fighting against injustice. As a child, Tamara felt terrified, indeed paralyzed, as she witnessed the repeated physical abuse of her younger sibling. The warrior game gives Tamara the opportunity to see herself as active and forceful, rather than frozen and frightened. In the game, Tamara becomes more of who she once could only wish to be. There is reparation. And in her conversations with me, she is able to rework this material with new perspective.

Joanie

Joanie, thirteen, is bright and articulate, but overweight and unhappy. She comes to me to talk about her problems with girlfriends and her unhappiness about her weight. She feels addicted to a particular computer fantasy game. She says she plays it "incessantly," often at times when she should be doing homework.

> J: I know I shouldn't do it, but I just can't resist.
> ML-W: Any sense of what is so irresistible about the game?
> J: I really like who I am in it. You know, I created a character. It's a fantasy game.
> ML-W: Can you tell me what your character is like?
> J: Yeah, sure. She's adventurous, kind of sassy, funny . . . also, a bit flirty . . . and kind of forward, socially.

Joanie's involvement with her made-up character helps her sustain an image of who she wants to be. The image is potent. It carries her through a year of steady weight loss, during which she is able to eat more sensibly and to exercise. Joanie is animated when she talks about the character and the game, more animated than when she speaks about anything else. In therapy we try to reconcile who she is in the game and who she is in daily life, a person Joanie calls "everyday Joanie." The character has an aggressive quality that "everyday Joanie" does not acknowledge but starts to own as she discusses her online self.[7]

In therapy, Joanie begins to talk about her sense of defeat. She sees female classmates who seem to have less of a struggle with their weight and male classmates who are not romantically interested in her. These topics, formerly taboo, are opened up in conversation about her online self, someone who would not have these problems.

In therapy, Joanie and I explore why she does not feel able to compete with her girlfriends and how she uses her weight to remove herself from the competition for male attention. Through her avatar and then conversation with me, Joanie is able to take feelings that otherwise might have been too humiliating to discuss and channel them toward increasing her will to change.[8] The combination of gaming and psychotherapy helps her regulate feelings that might have stood in the way of her further growth. Gradually, her depression lifts. She feels closer in the real to the person she plays in the game and plays the game less.[9]

Recently, several years after ending our therapeutic work, Joanie returned to see me. In trying to become closer to her ideal, she encountered limitations that she now wants to address as a more mature person. She wants to work further on her relationships with boys, still easier in the virtual than in the real.

In adolescence, young people try to align who they are now with who they wish to be, in other terms, with their ego ideal. They refine this ego ideal using their maturing capacities for self-observation. It is important that the ego ideal be realistic so that the adolescent can feel in sight of his or her goals and keep self-criticism at bay. In Joanie's case, fantasy games helped organize her aspirations.

For Joanie, the online experience works largely in her favor. But some adolescents may become complacent in their games, afraid to take the risks that come with face-to-face relationships.[10] Joanie and I talk about such fears when she comes back to psychotherapy. She describes not becoming who she wants to be with boys in the real world. Joanie is still drawn to online games to escape her intense feelings of disappointment. Our work continues.

Billy and Lawrence

Billy, sixteen, is intelligent, handsome, and kind. Although he has friends, he feels marginal, "behind in the game." It was he who asked his parents if he could see a therapist. He begins our first session with a description of his problem:

B: I feel like a fraud. I talk a good line, but I really feel like I don't know what's going on.

ML-W: Everywhere? Or are you talking about some particular place or with some particular people?

B: I guess it's just with kids, in general. I mean, I have my friends—but they're a lot like me.

ML-W: And what are they and you like?

B: Nice guys, basically. Not into the macho thing. And girls who like guys like us. But with everybody else, I act like I'm more cool than I am. You know, like I'm into the blood and guts stuff.

ML-W: Does it feel like you have to be into that stuff?

B: Hey. I'm a guy, you know?

ML-W: And?

B: And guys are supposed to be into being tough, right?

ML-W: Even in these politically correct times?

B (in an irritated voice): That's an adult thing. There's not much of that among the people I know. You're still a faggot or a girl if you're a guy who's not into sports or willing to fight.

ML-W: You sound kind of angry when you say that.

B: Look, I don't think much of this political correctness stuff. I feel like there's no room to just *be* anymore, you know? It's like someone is always telling you what's okay to be and what isn't.

Billy is having trouble coming to terms with what kind of man to be. He does not respect media messages about the macho ideal, but he senses that the "chill" demeanor he affects in front of his friends does not tell a complete story about who he is. After all, at night, usually with his male friends, he plays violent video games for hours on end. His lack of consistency makes him feel like a fraud. He cannot reconcile the several aspects of himself.

Billy's "nice guy" presentation enables him to continue something of his childhood persona. Billy has not caught up with his body, his changed physical appearance, and increased aggressiveness. When Billy and I talk about his gaming, he finds a way to own his new feelings: we come to see that taking the role of violent characters puts him in touch with his new adolescent aggressivity. We talk about how, in front of his friends, he is able to demonstrate his competitive urges and how this helps him to feel "more like a man."

When Billy and I talk about games, we have a context for discussing when aggression is needed in "real life." In our therapeutic work, Billy's "either/or" thinking (he is either the nice guy or the violent one), characteristic of childhood, gives way to "and" thinking, the synthetic thinking that is one of the achievements of adolescent development.

When adolescence comes around, it is not easy to be a man and be sweet. It is out of keeping with cultural gender stereotypes at a time when they are of the utmost importance. Billy's task is further complicated by the ever-present images of violence in media. But in video and computer games such as *Mortal Kombat* and the James Bond series, Billy sees and accepts his aggression and competition as appropriate to the game context. (James Bond needs to use guns to fight his foes.) By expressing his anger in the game, Billy feels that he "got it out"[11] and can visualize himself (like the characters in the games) as a person of strength and focus. The games become a ritual in which aggression feels codified and safe.[12]

Another patient, Lawrence, age twelve, serves as a counterpoint to Billy. For Lawrence, involvement with violent video games has not been constructive. In childhood he was diagnosed with pervasive developmental disorder. Years of psychotherapy have helped him become more socially appropriate and have greater impulse control, yet as an early adolescent he is still socially clumsy. I learn about Lawrence from his therapist, who consults with me about how to respond to fantasies he is having about her. Lawrence wants to unbutton his therapist's blouse and touch her breasts but worries that it may "not be okay" to have such thoughts. In talking to his therapist about his wishes, Lawrence refers to a James Bond video game to pose his question about what is appropriate. His therapist says:

> He asked whether girls and women liked to have their men be very forceful when the men wanted them in a sexual way . . . that he had been watching James Bond movies and playing a James Bond video game, and he knew that lots of women found James Bond to be very attractive . . . and that James Bond was violent but charming with his enemies and with the women that he wanted. It was striking to me that Lawrence seemed utterly sincere in this question—that he really did not seem to be able to distinguish between what goes on in real life and what goes on in the movies.

Unlike Billy, Lawrence is at risk, an example of someone with a diminished capacity to distinguish reality and fantasy.

Adolescents often experience their strong feelings in surges, with heightened experiences of marginality and loss, excitement and anger. Computer games provide a new way to calm themselves: the focus required by the games can organize them.[13] Adolescents play the games over and over again; their very repetition can be soothing, a counterpoint to inner turmoil.

Like all good play experiences, computer games can lead to feelings of efficacy that are critical to healthy development. And yet, vulnerable adolescents can be incited by the games, not to master their aggression but to act on it. In the study of adolescence and media, interactions are dynamic and need to be explored in fine detail. The contribution of clinicians to the literature on computer gaming will surely be to reinforce the point that any simple generalizations about "the impact of games on development" fall short of the mark.

Marsha H. Levy-Warren is a psychologist and psychoanalyst who is on the faculty and a supervisor at the NYU Postdoctoral Program in Psychoanalysis, and a Training and Supervising Psychoanalyst in the International Psychoanalytical Association.

CYBERPLACES

Kimberlyn Leary

From Freud's early hydraulic accounts of the drives and defenses to the American pragmatism of ego psychology, psychoanalysis has always been a culture-bound enterprise.[1] Here I consider a recent manifestation of this truth: how psychoanalytic space has come to intersect with cyberspace. Everyday immersion in cyberspace calls into question traditional ideas about what is public and what is private, what is on the surface and what is deep. In cyberspace, multiple realities and identities coexist and negotiate in new ways.[2] All of this brings new selves and subjectivities into the places where psychoanalysts do their work.

As this is happening, the analytic enterprise itself is exploring contradictions in many ideas that it once assumed to be unassailable. So, during the past twenty years, the view of the analyst as an authoritative source of truth about the working of the patient's mind has given way to the view that psychoanalytic understanding is intersubjective, growing out of the conversation between patient and analyst.[3]

Beyond this, recent innovations in technique have extended many analysts' expressive participation in sessions. Models of psychoanalytic practice are gradually shifting from the relatively abstemious to those that emphasize analytic "gratification and provision."[4] For some analysts, these new measures remain occasional adjuncts to standard technique as, for example, when there is a rupture in the treatment due to an extra-analytic crisis in the life of either patient or analyst. But for others, the turn away from traditional neutrality is fundamental.[5] Into this ongoing conversation come the practices of cyberspace: email, chat rooms, fantasy games, the World Wide Web, the range of Internet-enabled social media.[6]

In *Life on the Screen,* psychoanalytically trained sociologist Sherry Turkle drew attention to the computer as an evocative object for thinking about contemporary subjectivity. Turkle underscores how computational ideas alter traditional ways of thinking about surface and depth, unity and multiplicity, and the authority of stable meanings.[7] She makes the point that computational models and psychoanalysis have forged a new theoretical alliance: in each case, rule-based systems have become less important than the unfolding of emergent process. In both the psychoanalytic and the computational world, she argues, postmodern ways of thinking about theory have prepared the terrain for postmodern practices on the ground.

Here I extend Turkle's ideas to the consulting room, where technology gives patients new opportunities to think through self and subjectivity. The practitioner must be ready to face new challenges and embrace new possibilities that the computer culture brings to the place of analysis.

Chats and Games

Matthew, a patient in psychotherapy, spends his evening hours pursuing online relationships. He has been unhappy with his marriage for some time and had considered having an affair. In an apparent compromise, he reports his involvement with several women he met in a chat room. It is in the nature of the chat room to challenge traditional views of the self as bounded and stable. There, as in the proliferation of online worlds, when one enters (as a "chatter" or a "player"), identity is treated as a matter of self-presentation.

Matthew's conversations with the women in the chat room are animated with breathless, sexually titillating talk, but none progressed to the cybersex that appeared to be his goal. Over time, Matthew becomes frustrated with these relationships. He feels that these women are withholding something of themselves from him. Matthew has the very same complaint about his wife, but this fact scarcely interests him. Nor can he be concerned at this time with how his own aloofness and disingenuousness contributed to the problems he has had with women. Matthew is now considering entering a chat room as a woman, hoping to meet a lesbian who might seduce him. As Matthew discusses his intentions, he begins to recognize that he has always believed that women reserve their emotional intimacies for each other. It excites him to imagine being loved by a woman who believes that he is a woman. Matthew tells me: "That would be quite a trick."

Matthew is self-consciously constructing alternate selves in an effort to mitigate his unhappiness, believing that these alternates might be more successful than he in getting their needs met. At the same time, Matthew fully expects that it is he who will reap the benefits of their adventures. Matthew hopes to derive something "real" from an imaginary contact. That is his "trick."

In all this, Matthew draws on a familiar fantasy of self-transformation, but the Internet offers several unique opportunities for the elaboration of this fantasy. For example, the technology enables Matthew to instantiate multiple selves. He can configure himself in the (feminine) shape he hopes will yield the intimacy he craves. Matthew can "pass" as a woman in the chat room because his online self is created through language rather than, say, an exchange of digital photos or through video streaming.

When in our work together Matthew tells me about his online pursuits, he is conveying to me, a woman, his efforts to locate himself among women. It makes sense to assume that at least part of his message is intended for me. How do I, as a real, live woman, figure in this psychic cyberdrama? I learn later that Matthew's intention to become a woman online was born after he had seen me dining with a woman friend in a local restaurant. For Matthew, online or offline, it is the presence of a real person with autonomous interests, outside of his control, that he finds both threatening and also the object of longing.

Another patient, Melissa, a young undergraduate student, brings a related set of issues to her treatment. She tells me she has met someone in an online fantasy game, a "knight" who has begun to tenderly court her. He attends to her, their conversations are erotic, and she is smitten. Some weeks later, Melissa presses the knight to reveal himself. For her, their Internet relationship is no longer enough. She wants to meet him for coffee at a local bar. The knight responds by telling Melissa that he is all of fifteen years old, a high-school sophomore. Melissa is chagrined but in conflict. She has enjoyed what has taken place between them and she doesn't want it to stop. But since her knight is fifteen, is she doing anything wrong if they continue? The knight mounts his own challenge. He also wants their erotic talk to proceed. In one email, he reminds Melissa that he is still the person he was on the screen. Melissa hesitates and then types back: "You may be the same in the game, but you are now different in my head."

Melissa's comment that her knight is "now different in my head" is a recognition of the limits of virtuality. Once she knows the knight to be a fifteen-year-old, he can never be otherwise.[8] Most clinicians would not fault Melissa's comment for showing a lack of imagination but would find it a healthy adaptive response. She has come to an important realization, absent in much of the over-enthusiastic literature on cyberspace: the computer makes multiple selves possible—but only to a point. Melissa can live on the surface, but at a crucial moment, the need for depth returns. There is, after all, a person behind the exciting typing, and she has reason to believe that person is only fifteen.

Emails and the Analytic Couple

Some months ago, I woke up early one morning to get caught up on some back email before heading to my office. Coffee cup in hand, I watched my computer download the messages that had come in the night before.

One was from a patient I was seeing in psychotherapy. Morgan was a talented woman whose difficulties lay with her inability at times to maintain meaningful distinctions between herself and others. One result was that Morgan was not always certain how to make sense of her own mental states and often found it hard to identify what she was feeling. Morgan lived in a world of limited resources: one person's gain meant someone else (usually herself) would lose. In the treatment, Morgan struggled with feelings of competitive rage toward me and the envious wish to destroy those aspects of my life (e.g., marriage or career) she wanted for herself. However, when she did feel she had succeeded at spoiling what I had by denigrating it, she also panicked at the thought that my "loss" might now mean that I no longer had the ability to help her achieve the things she desired.

Morgan had emailed me before with questions or comments, typically after particularly difficult sessions, in what I had understood as her effort to shore up a feeling of remaining connected when our relationship felt especially tenuous to her. I wasn't entirely surprised to hear from Morgan. I had just returned from a two-day trip out of town that Morgan had resented. She had learned from a mutual acquaintance that my trip away was to deliver a paper at a conference.

The email was as follows:

Dr. Leary, my capricious shrink. You are the lamest thing going. The only thing that matters to you is your pathetic little writing endeavors. You don't have the heart to stay with your patients. You have only ambition, drive, and single-mindedness. You have no trouble clasping my check in those perfectly slender and inconspicuously manicured talons of yours without having been around for the most needful moments in my recent, unmomentous life. Right now, you are a pale substitute for my symptoms.

I wasn't sure how to respond to Morgan. It was not yet six a.m. Her email hit home, quite literally—Morgan effectively became a virtual presence interrogating me about my absence. If Morgan wished me to feel guilty about my being away she succeeded, as I took in the full measure of her pain and rage. At the same time, I was struck by the time/date stamp contained within Morgan's message: "*Right now,*" she had written, "you are a pale substitute for my symptoms." This left open the possibility that Morgan recognized that at some other time she might feel differently.

In the quiet of my early morning study, I was also free to be genuinely impressed by Morgan's indictment, for it was a very eloquent dressing-down, displaying Morgan's prodigious if often blocked skills as a writer. Since I had responded to each of Morgan's previous emails, I knew I would answer this one as well. But what to say? I wrote: "Morgan, I read your message. Well, you do write very well—I think you managed to capture some of your feelings about me very deftly. You have a way with words." I also told her I would look forward to seeing her later that day.

When Morgan arrived for her session, she was surprisingly calm and even reflective. Soon I learned why. "I wasn't even sure I had sent that email," she said, "until I got your reply. I thought that maybe I had only dreamed sending that message, or fantasized it. I guess it was for real." At first Morgan didn't remember much of the content of her email. "I didn't even keep a copy in my sent box." She did recall, however, my message to her, saying that she wrote well but couldn't really remember what *she* had written to me. "I think I said something about your fingers being talons. That's very funny because I've always liked your fingers. They are so long. I've always thought you probably played the piano and could make beautiful music with those fingers of yours." As the session went on, Morgan asked me to compare what she gradually recalled of her message with what I remembered of it. By the end of the session, Morgan had reconstructed most of the message. "I really said all that?" Then she added, "But I must also have meant it." At the end of the session, she asked me to forward her message back to her so that she could see with her own eyes what she wrote. I agreed to do so.

Clearly, this vignette can be approached in different ways. From one perspective, Morgan found a way to speak her mind without initially having to own her feelings. But as we spoke, Morgan was more able to claim her desire to "get to me" as she saw that I could receive her anger without either of us being harmed and with both of us able to go on.

Psychoanalytic Subjectivities

Psychoanalysis—a discipline popularly assumed to be preoccupied only with the past—shows itself to be remarkably sensitive to present-day contexts, including the fluidities and boundary crossings of cyberspace.[9] Irwin Hoffman and Owen Renik have specifically turned to a postmodern sensibility in their psychoanalytic writing. They challenge clinical practice in ways that lead to new metaphors resonant with those of the Internet, Web, and the subjectivity they make possible.

For example, Hoffman puts relational struggle at the center of clinical work.[10] Effective clinical work depends on the analyst's ability to fall into a spontaneous authenticity with his or her patient. Analysts try to subordinate personal needs in favor of their patients' interests but also expect that they will *fail* their patients in some unique fashion. A good analyst can deviate from his or her preferred stance. Indeed, it is this "throwing away the book" that leaves open a space for therapeutic growth.

Hoffman illustrates how this space can open by describing a clinical hour during which he and his patient become able to explore the meaning of a transaction only after Hoffman meets the patient's demand for immediate help. In this particular session, Hoffman spontaneously offered to call the patient's internist to secure Valium for her—a drug that, as a psychologist, he could not prescribe. For this patient, interpretation is possible only after her analyst is willing to do something for her that in this instance lies outside of his preferred assumptions about how analysts are normally helpful (i.e., by analyzing rather than prescribing, something he might have characterized as an enactment).

Analysts' subjectivities are highlighted as they are called upon to become personally responsive in ways that will be unique for each patient. Such a therapeutic moment cannot be explicitly invoked or instigated by the analyst. It is, to use the computational metaphor, emergent from an interactive and intersubjective context that cannot be predicted in advance.

Owen Renik also suggests that the emotional responses of the analyst can have therapeutic effect but argues that the analyst only becomes aware of his or her involvement after it is translated into direct or implicit action.[11] Renik argues that clinical accountability should include demonstrated links between analytic work and *patient experienced* therapeutic benefit. For Renik, the ideal of the anonymous analyst is both impossible and dangerous. Rather than neutralizing the field, it promotes idealization by assuming that, if the analyst's ideas were known, the patient would no longer be in a position to think for him or herself. Analytic progress follows from a collaboration in which both patient and analyst are autonomous. In collaboration, the analyst will depart from habitual ways of working and therefore bear a measure of discomfort, just as the patient is asked to do.[12]

Both Hoffman and Renik see the analytic situation as one that gives the patient an opportunity to work out nonlinear solutions to complex problems. Each advocates a therapeutic process that is emergent and provisional—one that constructs itself as it goes along. Here, clinical learning occurs through reciprocal and recursive exchanges of information, especially emotional and relational information.[13] These computational metaphors of the therapeutic process have little in common with the archaeological metaphor that Freud used to significant advantage during the first century of psychoanalytic thought.

Indeed, contemporary psychoanalysis conceptually resembles the architecture of the Web and other media technologies. Psychoanalytic knowledge takes shape in local contexts and in custom-tailored connections between two people and their subjectivities. This change in analytic ethos requires the analyst to provide more by way of involvement than was the case in times past.[14] Authenticity and relational connection are increasingly recognized as the *outcomes* of successful treatment rather than preconditions for analyzability.[15] The new ethos finds its metaphors in interactive storytelling on the Web where narratives are shared among users. They have no fixed beginning, middle, or end. Stories are traded back and forth, with each user contributing to the actions and characterizations that develop. The narratives have no one author; they are the product of multiple interacting subjectivities. Similarly the outcome of an analysis is a linked set of cognitive, affective, and interpersonal nodes. It grows from the layered interactions of the analytic relationship as it has been conducted over time.

In one sense, psychoanalysis remains as low-tech a venture as one could imagine. The "hardware," if you will, is a private space. The "software" consists of the emotional histories, hopes, and dreads that analyst and patient each bring to the treatment. Yet, technology changes what our patients bring to us and how we conceive of our relationships with them.[16] The very terms in which psychoanalysts describe their work have changed. The language of medicine and natural science has given way to that of virtuality and connectivity.

While technology is frequently implicated in the etiology of the existential ills for which patients need help (i.e., the lone hacker isolated in his room), psychoanalysts are now attentive to the ways in which technology can mediate analytic experience and even assist patient and analyst to participate in a meaningful exchange.

Cyberspace creates new classes of imagination and subjective activity. The question of what is "real" and what is "fantasy" as well as the question of when such a distinction should matter—always an ongoing preoccupation for psychoanalysis—has become newly relevant to the culture at large. The living legacy of psychoanalysis lies in exactly the capacity to find in the new a glimpse of the old and to locate in the strange something familiar.

Kimberlyn Leary is Director of Psychology and Psychology Training at Cambridge Health Alliance and Associate Professor of Psychology in Psychiatry, Harvard Medical School.

Through Fieldwork

THE INTERNAL CARDIAC DEFIBRILLATOR

Anne Pollock

Internal cardiac defibrillators (ICDs) are machines literally and figuratively "close to the heart." These devices are similar in form to pacemakers but closer in function to the external paddle defibrillators made iconic by television emergency rooms. Indeed, the Web site of one of the major producers of ICDs declares each device to be an "emergency room in the chest."[1] First developed in the 1980s,[2] by 2001 there were an estimated 100,000 Americans with this life-altering technology.[3] Implanted to monitor dangerous arrhythmias and automatically shock the heart into a regular rhythm, ICDs are designed to protect those at high risk for sudden cardiac arrest. They are meant to extend life, but they also change its management and meaning.

Having a machine inside you that periodically jolts you back to life brings up questions once raised only in science fiction and philosophical bioethics. How do the jolts of the ICD—traumatic biotechnological interventions—change the lives they seek to prolong? How do they change the deaths they attempt to postpone? Death for the ICD patient does not wait silently; it is foreshadowed with every shock.

I interviewed eleven ICD recipients who lived in diverse parts of the United States and who had a wide range of financial and educational backgrounds. I also interviewed two wives of ICD recipients. Despite their diversity, their stories had much in common.[4] The patients find it difficult if not impossible to communicate their experiences of the shocks and the intimations of their deaths. Unlike paddles in an emergency room, implanted defibrillators operate with no attending medical personnel. The patient is alone, isolated in fear and pain.[5] Most poignantly, they have been offered a technology where not choosing treatment is presented as tantamount to suicide.[6]

Death and Life

"I died and then. . . ." This is the peculiar grammar of stories told by people with ICDs. The internal firing of the ICD is painful and brings one back from death, a repeated boundary crossing that writes a new narrative of life and death. Making that boundary so traversable evokes feelings of confronting the uncanny in the sense that Freud wrote about it, something utterly novel yet known of old and long familiar.[7]

In his discussion of the uncanny, Freud writes that the "immortal soul" was the body's first double. Doubling has its roots in the desire not to die.[8] Many of those implanted with an ICD experience it as a new kind of body double, so it is not surprising that they regard it with the wonder we associate with the soul. For example, Joel, a fifty-eight-year-old Californian, says that when he contemplates his ICD it makes him afraid, but it also brings him to a new spirituality. Joel focuses on his good fortune at having survived his heart attack. It gave his life new meaning:

> I pulled out of it, and 97% of most people don't. So how lucky is that? And you have to deal with this stuff. Why am I the chosen person, out of those hundred persons—why am I one of the two or three chosen to survive? And I feel that there's been some—that something happened that I was chosen to survive. I have no idea what that means.

The darker side of having a body double is that the uncanny object brings death into new focus. Linda, fifty, a rural Southern woman, says:

> My independence was gone. And yet they say that this thing gives you *more* independence. Because you can be assured that you won't go into cardiac arrest and die when you take a trip and all that. My thing is, we take a trip, and I'm wondering, okay, I wonder which one of these exits is a hospital. Or, you know, something like that.

Initially, Linda was concerned about having a "foreign thing" in her body. That is something that no longer bothers her. Now she says that what is distressing is "knowing this is what I have to depend on. That I can't depend on my own body to keep me alive." Freud helps us understand the complexity of the ICD's promise of protection. He could have been writing about the ICD when he says of the soul as double that it shifts from being "an assurance of immortality" to "becom[ing] the uncanny harbinger of death."[9]

For patients who receive an ICD after a near fatal heart attack, the device is a reminder of both the death they escape and the one they will someday have. Joel puts it this way:

> Every time I look in the mirror I think, oh, you've got an ICD in your chest. There's a physical manifestation of what happened to me. It's something that happened inside my body, but I can see it every day when I take a shower. I look in the mirror and I see a little lump. Yeah, I think about what happened to me every day.[10]

For those patients who receive the implant after diagnostic tests indicated it was appropriate, shocks serve as a reminder of the death they will not have, a sudden death. A forty-two-year-old worker from the Rust Belt, Stan received his ICD when he passed out while running. Now he considers that the death he almost had would have been an "easy death." "Like blacking out on the road, dying like that would be nothing. There would be no pain whatsoever. . . . To black out that way and die would be the way everyone would want to go, almost." The ICD spared him that "easy death" and in the future would wake him up after a similar heart incident. In our conversation, Stan refers to an article on ICDs that we had both read in the *New York Times Magazine*.[11] Something in the article struck him: "Somebody mentioned in the article [that] it takes away the way you're going to die." Stan feels that the ICD has allowed him to make a trade-off. He gets, and is grateful for, the extra time: "I don't want to die tomorrow." But he has lost the easy death. His greatest fear is that he will receive multiple shocks from his ICD and then die.

Indeed, on one occasion, Stan did receive multiple shocks. He was swimming and felt a "funny feeling" in his chest that made him stop. "And all of a sudden, wham, I got shocked—damn, I gotta get out of the pool." Just as Stan was getting out of the water, he was shocked again. He tried to explain to the lifeguard what was going on. He gave her his medical necklace and pulled the card out of his wallet that told her who to call. Then he was shocked a third time.

After the incident in the pool Stan asked his doctor how many times the ICD would shock him before it "would stop trying." His doctor told him "about nine times." Stan struggles to articulate why he finds that number high:

> It will stop, reanalyze, go off, right, within about ten, fifteen seconds of each other, probably. Or maybe I don't know what the span is between shocks, I forget, when I come out of the pool. It

might have been twenty seconds, thirty seconds, maybe. Yeah, it will go off nine, even if it doesn't cure it, it'll keep going off. . . . I mean, it's like, if you're going to die, you're going to die. If you get shocked nine times, I don't know if, [sigh]. Yeah, it's supposed to correct it, but we'll see. Like, sooner or later, everybody has to end—know what I'm saying? I mean when I get way, way, way older, or whatever . . . and my number's up, I don't know if I'm going to get shocked nine times before I die, or some, I don't know. Or if I'm going to get shocked at all, if your heart stops, depends on how you die I guess.

It is a new thing to know the way that one will *not* die. I met Samuel, forty, in a café in one of the small cities that lie on Boston's periphery. He had received his implant only a few months before. He was a large man at 300 pounds and wore a grey sweat suit. Samuel's wife, Sarah, accompanied him. She pointed out the strangeness of understanding that "you'll never die of the fatal arrhythmia you've been diagnosed with." She asked, "How many people with a diagnosis can say that, that they'll never die of their diagnosis?"

The ICD preserves life, but can provoke a new agony of life painfully extended. The body becomes the machine's object, and that machine gains the power not only to save life but to terrorize it. In this, it again evokes Freud on the double: "The 'double' has become a thing of terror, just as, after the collapse of religion, the gods turned into demons."[12] The ICD offers the fantasy that death is avoidable. It turns each patient into an exemplar of the uncanny, where death is always close but never determined. Freud suggests:

Biology has not yet been able to decide whether death is the inevitable fate of every living being or whether it is only a regular but perhaps avoidable event in life. It is true that the statement "All men are mortal" is paraded in text-books of logic as an example of a general proposition; but no human being really grasps it, and our unconscious has as little use for it now as it ever had for the idea of its own mortality.[13]

With an awareness of their forestalled deaths, ICD patients develop magical thinking not only to ward off death, as do many of us, but to ward off the shocks themselves. Although the shocks are painful, patients say that what makes them almost unbearable is that they happen without any notice and with no discernable pattern. Stan describes the shocks as "aversion therapy" in the spirit of the movie *Clockwork Orange.*

ICD patients are under a machine surveillance that evokes historian Michel Foucault's description of the Panopticon as a prison with a guard at its center, making it possible for the prison guard to see the prisoners at all times. Indeed, the prisoners always feel under the guard's gaze, whether he is actually there or not.[14] The regime of control for an ICD patient is even more comprehensive than that for such prisoners. Foucault's prisoner needs to internalize the gaze of the guard. The ICD patient's surveillance does not need to be internalized; it begins by being within. Foucault's prisoner knows what constitutes transgression. The ICD patient does not know what actions will trigger a shock. Desperate to take some step, any step, to avert shocks, patients do things they have no reason to believe will protect them. Stan, for example, altered his beloved exercise routine ("I don't go all out like I used to") even though his ICD had never gone off while he was exercising.

Similarly, John, a seventy-two-year-old retired engineer from Montreal, tries to calm his excitable nature although the ICD has never fired while he has been agitated. His effort to control his destiny is all the more desperate since he is trying to placate a fallible machine. He has been repeatedly shocked by a faulty ICD:

I would be sitting quietly, and this may sound rather facetious, but anywhere from sitting, reading the newspaper, and all of a sudden this thing would go, bang, and I'd kick over the coffee table and say what the hell happened to me, to I'd be sitting on the john, and, bang—boy, I tell ya—a few of those and that would induce constipation. . . . [he laughed] So this occurred during July '94, it occurred all the various, various times—in fact I had two firings in one day. And when that happens, it can almost put you right over the hill. I think the body is tuned to react to electrical discharges, and it can become very frightening, disheartening, discouraging, what the hell's happening. And nobody could tell me.

John is in a situation where a course of rational action eludes him. He wants to avoid pain: "It's one of these things that after you've had a few of them you do your damnedest to avoid them." But the machine is faulty and can fire for no reason. He ends up feeling dehumanized, like an animal: "Like cattle with an electric fence around their field. After they brush against it a few times, they stay away." Yet from John's point of view, the cattle have an advantage over him: he doesn't know how to stay away from the fence. He wants to know the contours of the fence. All he has is second-guessing and magical thinking.

Before and After

The decision to receive an ICD does not fit easily into medicine's standard categories of informed consent. ICD patients are most likely to talk about their "decision" to get an ICD as no decision at all. When they consider how their doctors framed the decision, it recalls how social theorist Slavoj Žižek describes a forced choice. Žižek's examples of forced choice include the demands made by the United States on other countries when the United States supports elections only after an unacceptable group has fallen in popularity or has otherwise been removed from the ballot, thus granting others "the freedom to make a choice on condition that one makes the right choice."[15] Forced choice is an imperative masquerading as a choice.

Doctors present the ICD as the *right* choice. My informants say that their doctors present implantation in the context of an imminent threat of death. John says he chose the "obvious" when his doctor said: "Okay, you can die or you can have this thing." Since receiving his ICD, John has become less sure that his heart attack was necessarily explained by a heart defect. The life-threatening experience that led to the ICD occurred after a particularly stressful week, and he is convinced that all but one of the ICD "firings" (his word) that he has experienced were due to mechanical defects in the defibrillator. Moreover, John is not happy that the defibrillator was presented to him as wholly benign: "They said it won't do any harm. Obviously the guy who said that has never had one of these things fire."

John's choice was made without understanding what was at stake. Like other informants, he insists that it was impossible to imagine what life with an ICD would be like. The most resentful patients say that their physicians never acknowledged that the machine comes with a cost.

Linda was offered her ICD after a potentially fatal heart attack. She had been experiencing symptoms that her doctors ignored. Finally, on the day she was put on a heart monitor she went into ventricular fibrillation. She says that either this was a coincidence or "God talking." When Linda had her medical crisis, the doctors presented the choice: this machine or your life. She told me: "When the doctors look at you, and they say, 'Well, you know, if you didn't have this you'd be *dead*,' it's like 'okay, thank you.'" But no one ever told her what the experience of having her life repeatedly saved by a machine would be like. She has been shocked by her ICD over eighty times. She talks about it being a good thing that "they have it." The impersonal syntax is telling. Linda believes the ICD has saved her life, but the pain and uncertainty in her life make her unable to describe the good it does in personal language. It is good that "they" have it—that it exists in the world. When it comes to her own case, however, Linda is not so sure.

Once an ICD has been implanted, my informants feel that asking to have it removed means choosing their death. Samuel says: "You'd be committing suicide if you have what I have or what other people have and you take the boxes out." In Samuel's formulation, rejecting technology is identical to suicide. When sociologist Emile Durkheim wrote his classic work on suicide in 1897, he saw suicide as a discrete category. In contrast, the decision to remove an ICD can be an element in a set of choices for a better life.

Barry, a fifty-year-old information technology professional from the suburban East Coast, had an ICD implanted after a heart attack. Then, through online research, he learned that he was not an appropriate candidate for an ICD. The doctors had made a mistake. At first this knowledge made him nervous about his ICD misfiring, but in the seven years he has lived with his ICD it has never gone off. His first ICD ran out of battery power in 1999 and Barry considered not replacing it. But he decided that although the ICD should not have been implanted, now he wanted it. For him, it had come to represent what he calls "insurance." His reasoning: even if it wasn't put in for the right reasons, he is getting older (age fifty when he and I spoke) and since it was not shocking him, why not leave it in?[16]

Barry told me that his cardiologist likes it when Barry attends ICD support groups. The cardiologist said that Barry provided a "positive example." On the surface, it is absurd for a doctor to choose Barry as a role model. Barry does not need his ICD. He has never been shocked. His life is radically different from those patients who presumably need the support groups most (those who are shocked often or who have a poor prognosis for survival). What we see here is the physician's denial of the toll taken by implantation. The physician cannot see that his often-shocked patients cannot be optimally supported by one who has never had the experience. Yet considering this case can help the rest of us think through what ICDs may come to mean for many more of us: technological insurance that is hard to turn down.

When Barry agreed to have his ICD battery replaced, he did so because taking it out felt like a tiny step toward suicide. Barry accepts that he and the machine are now one; he has a cyborg identity. Samuel has a similar thought in this dialogue with his wife that I witnessed shortly after he received his ICD:

> Samuel: I've got something inside me that I know, forever, has changed my life.
> Sarah: I don't think there's even that option. You cannot be the old you.
> Samuel: I don't think I ever would return to the old person.
> Sarah: It's like me before and after kids. They call it a transition, it's not, it's a metamorphosis.
> Samuel: You can't get rid of kids either.

In general, ICD patients were glad to speak with me about their experience, but they stressed that they had been through something that was in its essence, incommunicable. Stan said:

> Nobody can understand what the feeling's like, to get these, it's kind of like getting electrocuted from the inside. Imagine that. When it drops you to your knees, you know you're getting hit pretty

hard . . . with electricity. And they [the doctors] can't understand that, and I would like to whack them with this so they would understand what is going on here. They're good at what they do, they're good at putting these things in probably, and they know they're saving lives, but they don't really understand completely what's going on with it. And I think they have a certain mindset [toward] the people that they're giving it to, yeah, there's going to be problems but that's just the way it is.

Literary theorist Elaine Scarry has argued that pain creates gaps in communication.[17] In its inexpressibility, pain tests the limits of language. The isolation of a person in pain is a central fact of ICD patients' existence. They want to talk to their doctors about their pain to develop physicians' empathy, but know that they cannot communicate what the shocks feel like. My informants were convinced that their physicians did not understand how their lives were affected by the shocks. As Samuel put it, "They [the doctors] think of you as a conduction defect."[18] His suggestion for the doctors: "I'd like to get those bastards and just shock them with 800 volts just to let them have an idea what it feels like."

Although the shock that patients call "the zap" and doctors refer to as "therapy" is not easily spoken about, it is what prospective patients and their families most want to know about.[19] The most frequently asked question on a Web site by and for ICD patients is "What does a shock feel like?" The Web site's answer tries to be honest while remaining upbeat. It asserts the necessity of having the ICD by comparing it to death.

That's a tough one. Like a sneeze, everyone's reaction is different. Some people describe it like being kicked by a mule, others hit by a two-by-four, still others describe the rush of electricity through their body to the ground. Some people black out and may collapse before "therapy" is administered. Others are conscious for the whole thing. The lucky ones feel a little tingle. Suffice it to say that it's not the most pleasant experience that you will encounter during your life. However, it sure beats the alternative!

All of my informants spoke about Dick Cheney. Cheney has an ICD but claims never to have been shocked. Some of my informants wonder if, on the contrary, he has been shocked and this explains what has become of him. They say that they recognize the fear and pain in his eyes, a certain look. Others believe his claims not to have experienced a shock and say that this is why he is able to continue working. They emphasize how much support and help he gets, trying to understand how an experience that is so debilitating for them could be shared by someone with such power and responsibility.

Some pundits have allowed themselves to imagine that it was in fact Cheney's ICD that turned him from the man who urged caution when he worked for the first President Bush to the man who promoted apocalyptic and rash policies under G. W. Bush. This point of view is captured by Maureen Dowd, who writes: "Some veterans of Bush I are so puzzled that they even look for a biological explanation, wondering if his two-year-old defibrillator might have made him more Hobbesian."[20]

References to Cheney portray a diffuse anxiety. Does his status on the boundary between life and death make him a threat to the living? What could this have to do with escalating global war? Internet sites and comic strips make half-jokes that Cheney is already dead, an undead malevolent cyborg. This would make him the epitome of Freud's uncanny and a channel for anxiety about a cyborg self.

Cyborgs

In the mid-1960s, cybernetician Norbert Wiener described the problem that doctors would face should it become technologically possible to prolong life indefinitely. Wiener cautioned that with the decline of quiet euthanasia, such as letting the too-frail die of pneumonia, doctors would engage in active euthanasia because families would not be able to bear their loved ones' suffering and societies would not be able to tolerate the cost of indefinite yet degraded life. In the new order, the doctor would become more god-like than before. Wiener wrote: "What if every patient comes to regard every doctor not only as his savior but his ultimate executioner? Can the doctor survive this power of good and evil that will be thrust upon him? Can mankind survive this new order of things?"[21]

Wiener's predictions have not come to pass, but they are wrong in an interesting way. Wiener imagined doctors making the life and death choices. But ICD patients illustrate a different scenario in which the choices are being left to each of us. ICD patients are harbingers of the time when we all will be asked to accept or refuse imperfect medical technologies, and accept the role of being our own "saviors and executioners."[22]

Stan explicitly weighed the question of whether life is worth living in the pain and fear that accompanies his ICD. Stan was not actively considering removing his device, but leaves the possibility open: "Somewhere along the line you gotta weigh 'what is the pain worth?' I don't know how many people out there got them, how they're dealing with it, but I bet you they're thinking similar thoughts."

What model of the cyborg, then, is provided by people with ICDs? Donna Haraway defines the cyborg as a "hybrid of machine and organism, a creature of social reality as well as a creature of fiction."[23] Some cyborgs seem triumphal (such as the pioneers of wearable computing), but people with internal defibrillators are cyborgs who show the strain. In this, they may be emblematic of the cyborg future in an increasingly geriatric North America. They do not have models to explain their pain or think about their isolation. They turn to their doctors, and their doctors are mute on what matters; they turn to articles in newspapers that offer more hype than help; they turn to online support groups for the simplest recommendations on how to live. Their experience reminds us that the machines we put in our bodies are as imperfect as our bodies themselves.

Anne Pollock completed her PhD in the History and Social Studies of Science and Technology at MIT in the spring of 2007. After a postdoctoral year in the Department of Anthropology at Rice University Department of Anthropology, she became Assistant Professor at Georgia Tech in the Department of Literature, Communication and Culture, contributing to that department's program in Science, Technology and Culture.

THE VISIBLE HUMAN

Rachel Prentice

The image on my computer screen depicts a woman
against a blue background. She has an upturned nose
and a double chin. Her eyes are closed and her wide
mouth is set in what could be a grimace or an enig-
matic smile, a Mona Lisa smile. Her brow is furrowed
in concentration, discomfort, or pain. Her body is rigid;
her arms are held tightly against her sides, and her
legs are pressed together. She has a paunchy belly and
wide hips. Her skin is a bronze color and an unnatural
light gives it a strange, shiny look, like plastic where it
is smooth, like hammered metal where its surface is
rough. Two lines, like raised scars or seams, run from
the top of her head, down the sides of her face, around
her breasts and abdomen, and down the center of each
leg. A scar—a deep gouge—marks the inside of her right
thigh. The woman has no hair.[1]

The Visible Human Female, as this digital creation is called, is a new object in the world: a digital reconstruction of a woman's corpse. Dead bodies in any
form can evoke strong emotions. But most North Americans see corpses in only one of three contexts: medical, funereal, and in films or other entertainment. But
the Visible Human Female and a similarly constructed
Visible Human Male do not belong in any of these contexts. They exist somewhere between real and artificial.
They are not dead bodies. They are not photographs of

dead bodies. They are digital reconstructions of actual dead bodies. They resemble photographs of people, but the images are strange. Color and shadow are subtly wrong. There are odd marks. They seem to have no context other than the computer screen. These images are uncanny; that is, they are simultaneously familiar and unfamiliar.[2] They look like bodies we have seen, but not quite. They are difficult to classify and yet not so horrifying that they override opportunities for thoughtful observation.

The image of the hairless woman is made from CT scans of a real woman, a fifty-nine-year-old homemaker from Maryland. She died of a heart attack, and her body, donated to science, wound up at the University of Colorado. There, researchers made the original cross-sectional images from which another research team would create the Visible Human animation. The woman was the second of two bodies imaged in this way. The first was the body of a death-row inmate, thirty-nine-year-old Joseph Paul Jernigan. Jernigan's name was tracked down by the press after project officials revealed that he had died by lethal injection. The woman's name was kept anonymous.[3]

The images of these bodies are part of the Visible Human Project, an effort sponsored by the U.S. National Library of Medicine to bring anatomy into the digital age. Project organizers wanted to create a digital archive of the human body that experts in many disciplines could use as a standard tool for research and teaching.[4] Anyone who wants to manipulate the databases of the cross-sectional images can access them for a small licensing fee. Thousands of licensees around the world are creating, among other applications, reconstructed images of bodies, simulations of body processes, cinematic bodies, and medical illustrations.

To make the image database for the visible woman, researchers first embalmed and then froze her body.[5]

They sectioned it using three technologies—magnetic resonance imaging, computerized tomography, and digital photography. MRI and CT scans image the body's interior without physical intervention. Digital photography is the only method that reveals light, shade, and color as the unaided human eye would see.[6] To make digital photographs, researchers cut her frozen body (and that of Jernigan) into four parts. They then used a special sanding machine to grind the frozen body into cross-sections corresponding exactly to the CT and MRI scans—one millimeter thick (the width of a sharp pencil tip) for the man and one-third of a millimeter for the woman. Researchers sectioned the woman more finely to achieve greater resolution. They took digital photographs of each shaved-off body section. The frozen shavings were later cremated.

Much press and popular attention focuses on this act, often describing the sectioning as "slicing" like a salami, rather than shaving. Articles dwell on the gruesomeness of the sectioning, even though cross-sectioning is a longstanding practice in anatomy.[7] This reveals one way that standard anatomical practice, as with many scientific practices, remains unremarkable when hidden within the laboratory but appears more sinister when it leaves the laboratory.

After imaging, researchers used a technique known as volumetric rendering, which gives the appearance of three-dimensional volume and depth. Neither a photograph nor a drawing of this woman's body in exactly this state was used to create this image. Instead, researchers digitally knit CT scans to create the figure of the woman's body, like building a loaf of bread from its slices. They also animated the image, so that, as the viewer's perspective proceeds down the woman's body, her skeleton pops out of her body and mirrors this progression.

What happens when virtual dissection moves out of the closed space of the anatomy laboratory onto the

Internet, where anyone can view cross-sections and re-constructions of these bodies, their skeletons, organs, and other parts? I asked twelve people to look at several Internet-based images of the two visible humans and interviewed them about the experience.[8] Not surpris-ingly, the images evoked strong reactions.

Here I focus on three interviews in which people responded to the visible woman by dwelling on the boundary between living person and dead body. These three people, two women and one man, explored this boundary by reflecting upon the circumstances that led to the creation of the digital object.[9] Each picked up on a moment in the visible woman's history when, for them, the relationship of "person" to body seemed es-pecially charged.

The three informants include a medical doctor with clinical experience, an art therapist, and a medical writer without formal medical training.[10] They range in age from twenty-seven to forty-six and, although none had experience with images from the Visible Human Project, their experiences with dead bodies ranged from gross anatomy classes and funerals to no encounters at all.[11] I began interviews by asking individuals whether they had ever seen a dead body before and, if so, in what context. I told them how the images were made. Then I showed them several images and asked, "What do you see?"

Wendell

Wendell, thirty-one, is a physician who trained in Hong Kong. For him, the visible woman raises some of the questions medical students face when they confront their first cadaver: What is this thing that once had a life but now is an object? Wendell and I begin our conver-sation by discussing his medical school education. He and his laboratory partners had to retrieve a wrapped body from storage and clean it before starting work:

The first sight of the cadaver was when we washed her. It was like a doll. It's a medical object, indeed, but it's not like a model. It's an object like a doll, but it's not perfect that way. But we washed it like a doll, like a thing. . . . When we say "medical object," it doesn't mean it's like a model or a mold. Everybody knew it was a dead person. It was a fact. You could not help thinking about that. But it was just there, an object waiting to be explored.

The cadaver Wendell worked on in medical school was hard and frozen. Its color and feel were unlike human skin. But the cadaver was not "just" a doll, model, or mold. A manmade model or doll would not have the marks on the body that a real body acquires during life. Wendell describes seeing a bullet hole in one cadaver and noting a missing uterus in another—connections to the life that the "medical object" once had. The cadaver was both a person and a thing.

In order to practice, physicians work out when it is appropriate to think about treating "the person" and when it is appropriate to treat "the body."[12] They use the distinction to distance themselves from anxieties about crossing body boundaries.[13] One surgeon I have worked with uses food metaphors to describe the feel of tissues, likening cartilage to coconut, for example. She describes this as "displacement," using the psychoanalytic term for replacing something highly charged with something less charged. The medical school cadaver has a role in this negotiation of affect. Medical students can bond with it as they might with a patient and yet treat it as a thing, an acknowledgment of the violence they will do in dissecting it. In Wendell's experience, the very use of the word "cadaver" helped him keep a distance: "Once you use this term, it seems to me you're not seeing the dead body."

Wendell's anatomy training included group dissection of a cadaver in a large laboratory. Wendell gave his cadaver a certain agency. He redefined it as a "work partner" who would help him learn anatomy. Wendell believes that the legacy of his medical school cadaver was that he learned to think of healing as a partnership between doctor and patient, something that later made him a better clinician.

When Wendell looks at the visible woman, he is kept in a medical context by his familiarity with the procedures used in the preparation of her body. He notes that the body is shaved and positioned according to standard medical practice. He notices the two lines running from the woman's head to toes—not part of common dissection practice—and wonders what they are. They are tubes that were filled with copper sulfate and glued to the body before it was imaged and sectioned. The tubes then appeared in cross-sectional images and helped researchers properly align the sections. In full-body reconstructions, these tubes tend to look like scars running from head to feet.

Wendell notes that the image of the visible woman has the same bronze color as the embalmed and frozen bodies he dissected in medical school in Hong Kong. Embalmed bodies lose the hues of living skin. In death, the back flattens out, losing its natural curve. Although Wendell says the visible woman looks similar to the cadaver he dissected in medical school, it does not evoke the same emotions: "I do not think of it as my work partner because there is no physical contact." He worries what will happen if students of the future, who might work only with virtual bodies, do not have an opportunity to form a bond with a cadaver as a "partner."

For Wendell, anatomy was a physical experience that taught him how to touch a body, how to work with a body, and how a body looks, feels, and resists when opened. In his view, opening a body should entail resistance. Without it, Wendell fears the loss of a certain sanctity. The double click of a computer mouse does not suffice to mark the moment.

> Dissecting a cadaver, you really feel it's very real. With this, the skin never serves as a boundary. You just click your mouse and it will disappear. There is a physical tension when you open up a [physical] body [here, he makes a gesture of prying a ribcage apart]. It's not there with the virtual.[14]

For medical students, the cadaver marks a passage into medical practice, where they still deal with a person's body as an object of inquiry.[15] The knowledge that the cadaver was once a "real" human being is critical to this transition. D. W. Winnicott described "transitional objects" as those that designate an "intermediate area of experience," typically between the infant's self and other things in the world.[16] In this sense, the cadaver is a transitional object, halfway between a model for practice and a living human being. Wendell asks us to reflect on how marks on the body, physical contact, and resistance to opening reinforce a cadaver's connection to the bodies of actual patients. Wendell fears that the visible human will not be able to serve this crucial role.

Cybil

Cybil, twenty-seven, is an artist and art therapist, who initially relates aesthetically to the images of the visible woman. Cybil looks at an animation that enables viewers to "fly" through the woman's skeleton, swooping through a tunnel of ribs. She notes that the bones are discolored, "Normal bones are beige, not gray." She says the lack of context surrounding the gray skeleton makes it meaningless to her: "Death has meaning; it gives meaning to life. With the lack of context, this is meaningless."

These intellectual associations—thoughts of color, metaphor, and meaning—change dramatically when

Cybil turns from the skeleton to the visible woman's exterior. She becomes very unhappy, anxious.

> Yuck! It's really bad, the color of her skin. I don't like it. It's like the plastic cloth they have on the dead body. It sparkles, it's so shiny. There are these strips that look like stockings. And her facial expression is so terrible. It's like she's in deep pain. . . . I have the feeling that she didn't want it. Look at her arms and how tight they are. It's as if she's holding something and she doesn't want to let it go. Does she look pleasant? No. It's a terrible picture.

Cybil observes many things that others also notice: the odd look of the visible woman's skin, as though it's stuffed in a suit; the woman's registration marks, which look to Cybil like the seams of stockings; the rigidity of the woman's arms. Cybil also reads the woman's facial expression as pain and wonders about its meaning. Cybil also thinks about the woman whose body became the visible woman, about her feelings and intentions.

For Cybil, the marks on the woman's body are signs of an interior state. What has this woman done to herself? What have others done to her? "Now, I look at it as a body and I say they haven't done anything good. She didn't want this." Cybil is concerned that if medical students only work with virtual bodies they will lose sight of the people who once inhabited the real body. Cybil says: "If you see the body as a machine or as an object, you don't see it as a human being. It will switch your way of perceiving the world."

The longer Cybil looks, the more she sees the unhappy woman behind the computer image. Cybil says she would be happier about the image if she thought it was just a computer animation or a plastic model.

This woman doesn't want to be there. It might be that she died in pain. But under these circumstances, they are presenting these body images as if they don't mean anything at all. . . . I wouldn't want it done to myself. There should be intimacy or privacy for each individual's life, death, or post-death life. Post-death life doesn't exist, except it does now.

I ask Cybil to elaborate on her striking phrase "post-death life," expecting her to say something about how images on the Internet are, in a sense, eternal. But Cybil is thinking about spiritual questions: "I imagine this woman lying in a cold, huge room on a table. I can imagine people taking pictures of her and this woman lays there with this very strong facial reaction. I find myself asking, where is her soul or spirit."

Cybil, a Muslim, is not religiously observant. She is surprised by her thoughts about the soul and its afterlife. But she believes in respect for the body and individual privacy. She says she understands the scientific need to cut up bodies but asks why researchers simulated the woman's exterior if the project's purpose is to teach the anatomy of body parts, "I don't understand the logic they use here to show her face and her body like this, putting it together like this." She suggests, as do several of my informants, that depicting the cadaver's exterior this way violates the dead woman's privacy.

Cybil explains that, in Muslim funereal traditions, the spirit remains in the body for forty days after death. After forty days, people gather and play sad music to send the spirit on its way. Although she is not a practicing Muslim, Cybil sympathizes with these traditions. If there's any truth to them, the visible woman must be in agony: "There is pain. These things are brought together. The cuts. It really looks like they hurt her, that this is some plastic surface that she wants to get out of, that her soul wants to get out of. My God, this woman is dead. But her soul cannot leave."

In a two-hour interview, Cybil has moved from dispassionate analysis about artistic intentions to profound discomfort. The transition is fueled by Cybil's belief that the woman's exterior provides a view of a woman beneath, a woman in pain. The simulation cannot hold, and questions about the afterlife and soul become relevant. Our interview turns to the ethics of how the woman's body was treated. Cybil knows that the woman died of a heart attack. But what kind of permission did she give to be cut up as she was, displayed on the Internet indefinitely, with her face and its pain, her body and its identifying marks, shown to the world? Concerns about the ethics of making the man and woman visible ("did they consent to *this* use of their bodies?") were nearly universal among my informants.

Many cultures, including Muslim cultures, practice the double funeral. They hold two ceremonies. The first marks the event of death; the second, which usually occurs after the body decomposes, marks the soul's passage to its final resting place. The transition from death to decomposition is fraught with feeling. The body retains too much personhood to be thought of as mere remains. The double funeral gives a community time to grieve and, finally, to let go.[17]

Like cadaver dissection, the double funeral respects that the recently dead or preserved body is inanimate but inseparable from the person who once was. Only later, when the body has decomposed or been disassembled by medical students, does this connection to personhood recede. For Cybil, the visible woman has died, but the person appears trapped in her body. Cybil worries that researchers' treatment of the woman's body has neglected her wishes and possibly her soul. The woman has been unable to successfully make the transition from living person to dead body. Her soul, the essence of her personhood, is caught between living person and dead body. Her soul, the essence of her personhood, cannot leave.

Julie

Unlike my other informants, Julie, a forty-six-year-old medical writer, has heard about the Visible Human Project before our interview but is unfamiliar with the images themselves or how they had been produced. She first tells me she has never seen a dead body. She expresses concern about how she will react. She fears that she will identify with the visible humans: "It's difficult to separate the fact of deadness from assumptions about who that person might have been, issues that aren't very comfortable ones to confront. . . . The association will come back to you and the thought of yourself being on the table."

As Julie peers at the screen, she notes that the visible woman looks "intensely uncomfortable," but Julie is soon overtaken by curiosity about technical details: "I'm very curious about what look like scar lines [registration marks] and why she was sectioned this way. . . . I wish it loaded faster."

Julie is impatient with the download speed and not particularly interested in the image. She does spend time, however, trying to figure out where the woman's look of discomfort might have come from. Was it natural, from rigor mortis? Was it from the sectioning procedure? Julie begins by being matter of fact about the constructed nature of the image before her: "This is clearly a computer image. You can see the pixels. It's not an especially good one. It looks pretty crude." She says that she has no sense of the person behind the image, but as she focuses on the woman's face, her response changes: "When you look at the face, a sense of a real person having been behind it begins creeping in. In the expression, there's something that connotes . . . a human being. The eyes look painfully closed. There's a furrowed brow."

When I ask Julie where she believes these expressions of pain come from, she chooses a point after death but before imaging, a time when the body had entered a "liminal" state, that is, a transitional time out of the normal cycle of life and death.[18] She also takes it as a given that the woman had a funeral with a casket.

> My guess is this is not how she looked when she died. I wonder what changed as a result of the manipulations. She's been cut apart, touched, sawed, stored, moved around. How many hands have been on this face? I wonder how was the skin pushed here or there or whether it was touched at all. If I had to guess, I would say it [the pained look] was as a result of her being prepared for this project, or whatever changes happened before her funeral, in the casket, before it was decided she would become the visible woman.

Julie's remark about the funeral mystifies me until she makes a remark at the end of the interview, after I have closed my notebook. Although Julie had said that she had never seen a dead body, at the end of the interview she remembers that she has seen three. She tells me that she forgot about seeing her dead grandparents in their caskets before their funerals because she had placed my question about seeing dead bodies in a strictly medical context. Julie then goes on to describe the funeral of a grandmother with whom she was close. Before the service, Julie noted that her grandmother's lipstick was awry. Julie made the funeral directors fix it because in life her grandmother was very particular about her lipstick; improperly applied lipstick would have upset her grandmother.

On the boundary between life and death, the visible woman stirs a long-forgotten memory. Though she knew that her grandmother was dead, the improperly applied lipstick forced Julie to confront all that attends death: the body lying in front of her no longer had agency. It was at the mercy of strangers. Julie's speculation about the visible woman's trauma and casket calls forth memories of the moment when the body of her grandmother no longer held the living person.

Many objects of current scientific and technological practice, from computers that seem to have human-like intelligence[19] to brain-dead cadavers, challenge the boundaries of what counts as alive and human.[20] These artifacts either become familiar and lose their capacity to promote reflection or they remain within carefully circumscribed contexts where most people do not encounter them. In contrast, the visible humans, widely available on the Internet, retain their capacity to shock. For one young woman, a cross-sectional view of the virtual human resembles a butcher's cut through a chunk of red meat, "man as rump roast. . . . It makes us look no more unique than a cow." Jokes and odd metaphors may reveal unspoken anxieties.[21]

Julia Kristeva describes as "abject" those things that upset the categories that define our lives.[22] Corpses are abject, she says, not least because they disturb the boundaries between person and thing. Cadaver dissection utilizes the corpse's abjection, giving medical students an object that helps them to reflect on patient as person and body as object. The visible humans show us a dead body to which we have no personal connection, allowing explorations of the cadaver's unique ability to disturb the boundary of living person and dead body, leading to personal reflections, such as Julie's memory of her grandmother's lipstick, and to ethical questions, such as how to respect the wishes of the dead. In a world that shelters us from all but the most carefully contextualized death, such explorations can enrich our lives.

Rachel Prentice is an Assistant Professor in the Department of Science & Technology Studies at Cornell University. Her work focuses on technologies and bodies in medicine.

SLASHDOT.ORG

Anita Say Chan

Founded in 1997 by a self-described geek and Linux fan, the Web site Slashdot.org is currently a daily list of about twenty news stories, all focused on science and technology. On average, there are 1.1 million unique visitors each month, roughly 330,000 each day. Each story is contributed by a Slashdot user and topics range from Linux and the politics of free or open-source software to astrophysics, robots, and animé. Stories are linked to a user-moderated discussion forum where users can post their comments, criticisms, and add more links to the topic of the story. Typically, within hours after a story is posted, there are several hundred comments on its discussion forum: an August 2006 visit to Slashdot turned up a story about digital surveillance that had 439 comments posted to its discussion forum by the end of its first day on the site; a story about digital piracy's impact on PC gaming had 584 comments. Each user is able to rate other users' posts and users may then choose to view only the posts that the community as a whole has found most relevant. The experience of "news" on Slashdot is dynamic and participatory. Slashdot users say that they feel like students, reporters, editors, and critics. What is striking is that some also say that they feel like "addicts."

"Everything that I was interested in and everything that was important to me was on that Web site," says Patrick,[1] a twenty-three-year-old computer programmer. I meet Patrick when he has been a member of the Slashdot community for four years. He describes his introduction to Slashdot as a fall into addiction:

> When I was in college, I probably visited that site at least thirty times a day. I was reloading it every ten minutes. . . . Whenever I had access to a computer, I would check Slashdot. I would check Slashdot before I would check my email . . . That was the only Web site that I wanted to go to—I had my hotmail account and Slashdot and that was literally all I needed.

During his sophomore year, Patrick changed majors from physics to computer science. The move brought with it a forced migration from computers that used Microsoft Windows to ones that used the open-source operating system Linux: "Suddenly everyone was talking about Linux around me, and I didn't even know where to begin. . . . And I literally felt like an idiot." When he asked a fellow-student for help, Patrick was directed to Slashdot. He first used Slashdot as what he calls a "study guide," an archive and source of information to increase his understanding of Linux. Within a few months, Slashdot was more a constant companion than an archive. Patrick checked Slashdot headlines the moment he woke up and stayed up late at night reading and posting to its discussion forums. Four years later, out of college and with a job in programming, Patrick is on Slashdot three or four times a day. He leaves it running all night.

When Patrick talks about Slashdot as addictive, is he suggesting that giving and receiving information can cause harm, that news and conversation about it can be risky business? He is. Other Slashdot users also describe themselves as addicts; their collective use of the term is part of a larger trend that describes a wide range of human activities as the products of addiction.

Traditionally, saying that someone is addicted required the presence of an outside substance that could act on the biochemical makeup of an individual, causing the body to need more of it. These days, a range of activities (from overeating to the refusal to eat, from sustained lethargy to relentless exercising) are marked as having an addictive potential.

In this way of thinking, activities rather than substances are treated as addictive. As social theorist Eve Sedgwick puts it, blame shifts from a substance to a self, an addiction-prone self.[2] Addictive activity is any action that the addict himself cannot moderate. In these terms, by opting *not* to move off Slashdot, members of that community show an inability "to freely choose health."[3] And yet, although they use the value-laden language of addiction, those who claim to be addicted to Slashdot are not discontent. They speak of their condition as part of their claim to community membership. It is, in the main, a positive identity.

Learning Addicts

Mark is a twenty-four-year-old greenhouse technician in Raleigh, North Carolina, and an avid Slashdot user. Diagnosed with attention deficit disorder before entering college, he occasionally took Ritalin to manage it. He extols Slashdot for empowering users: "The *users* are the editors—they are the main editors." He adds that Slashdot employees "just post the stories—everything else is user-driven." Even as Mark describes his empowerment, he describes losing the ability to control the amount of time he spent on Slashdot. His description of his addiction as the "cultivation" of something "ridiculous" reveals his ambivalence:

> It was ridiculous for me—it affected my way of life. It was ridiculous how often I would go to the library just to load up a page on Slashdot. There

was a time when it was really bad. . . . I would just read that news site for hours and hours on end rather than do my homework because it was more interesting to me and I learned a lot when I was on it.

Mark recalls evenings when he would delay doing homework or going to sleep to spend several more hours reading through the news on Slashdot. "You have a week to turn in your assignment, so you can just put it off until the last minute. . . . So I could come home from class, and I just ended up refreshing Slashdot all night."

Mark knows that Slashdot distracted him from coursework, but it also gave him a positive view of himself as a learner. Indeed, it offered an alternative kind of learning that he felt was better suited to him than class-assigned textbooks: "Slashdot's probably more *in my nature* than a traditional textbook because there are lots of ideas in a short period of time and in a small space there that a user can read, rather than being something just concentrated on one subject." Mark believes that Slashdot not only helped him learn in a style more natural to him, it enabled him to be more reflective about himself.

Thomas, a twenty-four-year-old computer programmer in Manchester, England, describes a similar experience of cultivating a Slashdot addiction. Reflecting back on what he thinks of as his "obsessive" use of the site as an undergraduate computer science major, he says:

I didn't have a net connection at home, so I would go off to one of the labs in the morning. And as soon as I got in, I would have to pull up the site to see what had been updated overnight. Then, during the day whenever I had access to a computer—in between classes, in classes—I would open it up to see what had been updated [on the site]—and spend ninety minutes going through whatever had been changed. It was always the first thing to get read before anything else when I sat down at a computer—before email or anything.

Thomas read Slashdot during classes and, not surprisingly, those were the classes in which he under-performed:

> You would spend so much time reading the comments that it could take you away from school work. Especially if I was reading it in a lab, it was something I could do instead of programming. You'd open it up thinking you would only spend five minutes on the site to check it quickly, and end up spending forty-five minutes on it because you found one interesting story.

Yet Thomas is unapologetic about having used Slashdot in the classroom:

> I don't think it was that the classes were boring. I think it was that Slashdot just was so much more interesting. It was just so new and different to anything else. You kind of got the lecturer giving you information and standing in the front of the room and talking, and that's school and college—versus [on Slashdot, there was] this whole new world to play with.

That world is a world for learning. As Thomas sees it, a lecturer is saying what he already knows; Slashdot is about things newly discovered. The university professor is one voice speaking to many; Slashdot is about the voices of the many. Thomas insists that it was the social life on Slashdot that brought him into his current career, programming in the computer language Perl: "Just over a period of time, [issues around programming in

Perl] regularly came up [on Slashdot]. Like you would see people debating between the uses of Python and Perl, and you wanted to be part of the conversation . . . I would not have come across it anywhere else. And now I code with Perl every day."

Mark, too, sees Slashdot as crucial to his professional identity, but in a rather different sense. For him, Slashdot is not where he learned to do his job, but where he keeps his hopes alive for a better job in the future. An engineer, Mark can explore other professional identities on Slashdot: "It's kind of a professional thing for who I envision myself to be. Like I'd love to work as an [Internet protocol] administrator or Unix networking administrator. . . . [Visiting Slashdot] is my way of discussing these things with people I don't see day to day." His current colleagues see him as someone who can fix their email. On Slashdot he finds a more exalted community.

Now that they are working, Mark and Thomas have less time for Slashdot, but both say that they continue to feel "addicted." So, for example, Thomas recalls that he became frantic on a recent weekend visit to his parents' home, where he had no Internet connection and no "daily fix" of Slashdot. Both see their addiction as a response to a life disappointment: a failed educational system, a dead-end job. For them, addiction is not a state to be diagnosed but is itself diagnostic; addiction leads them to social criticism.

Working Addicts

Twenty-six-year-old Joseph is a computer programmer in Belgium and the author of a Perl program that sends updated Slashdot headlines to his cell phone every fifteen minutes. He created a Web page to distribute the code at no cost to other interested users. His page declares its users addicted: "Welcome, Slashdot addicted person! On this site you will find a little Perl script I wrote that sends the latest Slashdot headlines straight to your cell phone! No more reloading the main page all day, waiting for fresh stories to appear!"[4]

Joseph wrote his cell phone program at the height of his Slashdot addiction. He remembers feeling that, "If you didn't check Slashdot for a few days, you just felt when you did go back, you had to read everything." For Joseph, reading Slashdot is like watching an unfolding drama. "You just don't want to miss an episode. And if you do, you're afraid you can't follow something after all. So [on Slashdot] you want to see how news stories develop and what happened—what happened in the Microsoft trial or with a new bug that was discovered." When Joseph wrote his headline program, his habit was to refresh the site every five minutes to see if any new stories had been posted. With his code, it took less effort to feed his addiction: "I didn't have to waste time reloading the page [to see if new stories had been posted] because the code was doing it for me. . . . I could just continue reading or working or I could be in a meeting—I just got a text message and that was it."

When I asked Joseph how Slashdot affected his work, he did not talk about its impact on his actual job, but on how it stood between him and the job he might be doing. "It was almost surprising that no one seemed to notice [how much time I spent reading the site]. But you could say in a sense it was cutting down my productivity, because I could have been experimenting with stuff and improving them a bit a more. But everything was working fine." As Joseph began to think of Slashdot as an addiction, he began to think about his unrealized potential. He was functioning "fine" at his real job, but falling down on his idealized one.

Aaron, thirty-one, a technical project leader at a Minneapolis software company, admits that he is often late to work because he begins his day with Slashdot. Although Slashdot has gotten Aaron into trouble at work, he is proud that he has managed to integrate the site into his office routines and notion of productive work. He insists that while Slashdot might sometimes be disruptive, it is also a resource for problem solving. "Sometimes, for example, if I'm making the case for the use of certain kinds of technical tools at work, there will be posts on the site from people who have already made that case. Or there will be some new language or way at looking at programming." When Aaron says that he finds himself spending too much time on the site, he quickly adds that even if he can't always see a direct work connection, there may well be one. For example, he says that sometimes, the best solution to a problem is "distracting yourself—the answer will come to you." In this case, Slashdot helps at work by providing a distraction. The pattern we saw with the "learning addicts" is repeated here in a work context. Aaron's confessed addiction is not something he wants to control. For him, Slashdot supports new thinking about what is productive and nonproductive. It provides new insights, for example, that to be most constructive he needs downtime to think.

Dissenting Addicts

In June 2002, Thomas, the English computer programmer, read a story on Slashdot that upset him. It was titled "UK Government Expands Spying Powers" and included links to several articles in *The Guardian* about the British government's plans to expand a bill to increase the online surveillance of citizens.[5] Thomas decided to register his dissent to the new bill. He posted a dissent on Slashdot and faxed his letter against the bill to his Member of Parliament. Thomas says that he didn't consider himself politically active before he began reading Slashdot. It was Slashdot that got him involved with politics. He says: "The political side of things I would have had no interest in before I started seeing [political issues] on Slashdot. . . . When I started reading, it was totally a technological thing; I was just fascinated by computers and really not in other matters."

On Slashdot, Thomas became more aware of the importance of technology policy: "Now I know what the different parties do and where they stand on issues, whereas before I would have taken at face value what they said." Thomas even credits Slashdot for convincing him to vote: "I voted in the last elections [based on what I learned], and I don't think I would have without the political influence of Slashdot." Now, for Thomas, engagement with technology implies engagement with politics. His experience illustrates that although Slashdot has never formulated a political platform, voiced support for established political parties, or identified politics as a primary content category, people on the site begin to see technology and politics as related. Traditional modes of thinking about civic education would not include Slashdot as a force, but its addicts see it as a place to discover oneself as a political actor.

Danger and Addiction

In her history of addiction, Eve Sedgwick recounts a nineteenth-century shift in cultural perceptions when drug "users" were reframed as drug "addicts."[6] Unlike users, addicts were no longer assumed to have agency that they could exert with responsibility or authority. They were thrust into the care of legal and medical institutions. The addiction narrative insisted that only through such intervention could addicts understand and gain control over the self. A similar construction of the debilitated addict emerges in literature on "Internet addiction." Written in the genre of self-help books, their authors target their texts not just to the "addicts," but to the family members and loved ones who should presumably manage and control addicts' compulsions.[7]

Such warnings about individuals with overdeveloped relationships to computing are not new. In 1976,

MIT computer scientist Joseph Weizenbaum warned about an emerging category of compulsive programmers.[8] "Oblivious to their bodies" and "transfixed by their computer screens," they could be identified by their "sunken glowing eyes."[9] To Weizenbaum, they seemed to exist "only through and for their computers."[10] Weizenbaum sees this pathology as having significant similarities to compulsive gambling. Each has the driving force of "megalomaniac fantasies of omnipotence."[11] With the growth of computer technology, such depictions of computer enthusiasts became more widespread. For example, sociologist Paul Taylor describes the trend of hyper-criminalizing computer hackers. He points to what he sees as the excessive legal penalties and jail sentences for even minor infractions. In his view, sensationalized media accounts represent hackers as a threat to public safety. The media portrays them as "malicious computer geeks in darkened rooms, obsessively typing away"; as wayward actors disconnected from the morals, ethics and norms of the everyday world; and as "electronic vandals" who are able to do magical things with their computers.[12] Or, as Andrew Ross argues, hackers—like dropout students of the 1960s and the punks of the 1970s—were the 1980s' public example of moral maladjustment. For Ross, they are a test case for establishing control, for redefining and redisciplining the "dominant ethics in an advanced technocratic society. . . . [Today] the technology of hacking and guerilla warfare occupies a similar place in the countercultural fantasy as the Molotov cocktail once did."[13] In the terms that Mary Douglas uses in *Purity and Danger,* the subversive hacker is the "impure."[14]

For Douglas, classifications such as what is cleanliness and dirt, pure and impure, delineate the boundaries between what is safe and dangerous, between what is sanctionable and what must be condemned. Culture needs and generates such distinctions to impose order on a naturally disordered world. We know the desirable and proper when we know the undesirable and danger-ous.[15] Internet addiction defines the cultural boundaries of normalcy when it is interpreted as disordering pollution. It demands containment. Containment is only assured in the moment that addicts come to see themselves through the eyes of nonaddicts: as carrying a pollution that needs to be undone. Such containment need not be institutional. The genius of modernity is that in it individuals are taught to assess and contain *themselves,* a point underscored in the work of Michel Foucault, who describes how medical, legal, and political institutions make individuals aware of what constitutes transgression and that they are being continually assessed.[16]

For the purposes of this essay, Foucault's key insight is that, to manage addicts, they must be made legible to themselves as addicts. On Slashdot, we see users learning to read themselves as addicts, constructing an interpretation of addiction that makes them happy to continue. They talk about their addictions as enhancing their ability to learn, work, and act politically. A controversy between users and editors on Slashdot gives eloquent testimony to how this community perceived the *value* of addiction on the site.

Defending Addiction

In early March 2002, the editors of Slashdot posted an announcement that introduced readers to a new subscription policy.[17] From this point on, subscribers to the site could see and read pages without banner ads, while nonsubscribers could expect to see the advertisements incorporated into the content they would otherwise still be able to access without charge. The heaviest users would need to pay the most to see all of their pages without advertising content.

The announcement generated over 2,000 comments, the large majority opposed to the new subscription policy. Users objected that the proposal not only undervalued but *penalized* its most devoted contributors. There were reflections on how faithfully the most active followed Slashdot rules, how hard they worked, how much they had invested in a model of good Slashdot "citizenship."[18] The site's administrators insisted that only 18 percent of users would have heavy taxing for free pages, but those who commented on the proposal pointed out that "This site caters to the hardcore sorts that participate the most and are likely to fall into the 18 percent that have to worry."

Thomas was among those who posted a reaction to the proposal. He wrote, "So the people who made the site worth reading (the people who comment lots), are now going to be charged more than the people who don't give anything of value to the site? . . . Sorry but I just don't see how charging people who are content producers as well as the heavy content consumers is going to help the site."

Commenting on his post, Thomas tells me, "When you get the kind of people who are consistently posting, they're the ones who generally give more information than the story itself did, which is usually just some article written for a generalist audience." Who are the consistent posters? "If anything," he says, "they're probably the most boring people in the world—the people who spend their days on Slashdot. . . . I guess I kind of considered myself among them, when I was most active."

In Thomas's belief that Slashdot's heaviest users are both its most boring and most valuable, we hear his ambivalence about Slashdot addiction. Through one lens, the Slashdot addict can be depreciated as under-socialized, self-isolating, and "boring." Through another, the Slashdot addict appears as a valuable, contributing member of the community.

Other addicts express the ambivalence at the heart of Thomas's dual characterization of the addicted Slashdot user, pleased by what they are accomplishing but anxious that they may have crossed over to the "wrong" side. Aaron, for example, speaks of himself as using Slashdot "more often than I should," suggesting that there exists an ideal, balanced use of the site that has eluded him. And it is with a notable sense of alarm that Thomas describes his "slipping back into his old habits" while on vacation: "I spent several hours a day poring over it [during vacation], so if I had the time, I still might be using it that much. I didn't expect that I would be using it that much. I just thought I would go on in the morning and check it and then a couple of hours later, I'm like, 'Oh, it's eleven in the morning, and I still haven't gotten dressed!'"

Slashdot addicts do not celebrate their addiction in any simple sense. They see their behavior as something with destructive power, as something that requires vigilance. Yet, their addiction generates productivity as well as pleasure. It is most often the case that the same person sees the positive and negative aspects of his addiction.[19] Because Slashdot addicts experience the positive side of this ambivalence, they step back from "treating" their addiction. The term addiction is a contested terrain when addiction to a site on the Internet leads its "victims" to social criticism, educational fulfillment, political empowerment, and a sense of greater self-knowledge.

Anita Say Chan is a doctoral candidate in the Program in Science, Technology, and Society at MIT. Her research involves a study of electronic governance projects as models for development in Peru.

THE DIALYSIS MACHINE

Aslihan Sanal

It is not always true that patients want to survive organ failure no matter what the consequences. Dialysis and organ transplants alter an individual's sense of physical and psychological being; in terms used by the French psychoanalyst Jacques Lacan, it is a turn of comprehension.[1] The day we start imagining the inside of our bodies is the day we start living a skinless life, a phantasmagoric life. What is inside, impenetrable and solitary, magical and personal, becomes ordinary, visible, and vulnerable. The patient with kidney failure does not only experience the loss of bodily integrity but also the loss of sovereignty. Such a complicated life—from its beginning at the dialysis machine to the internalization, the literal in-corporation, of another body's tissue—might or might not be experienced as desirable.

Zehra, a young Turkish woman, and Oguz, a young Turkish man, both had kidney transplants in the same hospital in Istanbul.[2] Their histories and hopes are very different, but they have something in common: they both experience a change that seems secret and inevitable. To each of them, dialysis and organ transplants seem more than a "mere" medical prosthetic or an exchange of "parts." They seem a metamorphosis.

Zehra

She could be herself, by herself. And that was what now she often felt the need of—to think;

well, not even to think. To be silent; to be alone. All the being and the doing, expansive, glittering, vocal, evaporated; and one shrunk, with a sense of solemnity, to being oneself, a wedge-shape core of darkness, something invisible to others. . . . When life sank down for a moment, the range of experience seemed limitless. . . . Beneath it is all dark, it is all spreading, it is unfathomably deep; but now and again, we rise to the surface and that is what you see us by.

—Virginia Woolf, *To the Lighthouse*

Twenty-three-year-old Zehra's kidneys failed in 1997, when she was in Kocamustafapasa, a town close to the city of Kastanomu in northwest Turkey. She felt weak and sleepy, her body started swelling, and a pain slowly grew in her back. These were the symptoms from which her uncle Talat had suffered when he fell ill of renal failure. He went back and forth to Istanbul, he was on and off dialysis, but then, just as everyone thought he was getting better, he died in a bus on his way to the city for another treatment. So her parents took Zehra to a hospital in Istanbul when she started showing similar symptoms.

Zehra had to stay in the hospital for two weeks. The doctors ran many tests and told her that she would have to begin dialysis. Then, they opened a *whole,* a dialysis membrane, on her left fist. The first time she heard that she would need dialysis, Zehra cried. It could not be a nice thing to live on dialysis—three-hour sessions, three to four times a week. She had hoped for a long-term treatment, not being hooked up to a machine for the rest of her life. But the doctors gave her no other choice. On dialysis, she remained very ill: she was vomiting, her blood pressure was imbalanced, and she had pain in her ankles, neck, and joints. She could no longer walk properly. Painkillers were only a temporary solution.

On dialysis, Zehra felt as if her body was no longer her own, as if the body she had once known as her own was slowly being replaced in the cycles of dialysis. She disliked the blue dialysis machines, which had fluids different from those used in newer "white" machines. After six months of suffering with the blue machine and low blood pressure, she was switched to the white machine. To avoid interaction with other patients, Zehra tried to sleep during dialysis. Most of the time, she dreamt of herself on dialysis. In this recursive dream, the needles would come out suddenly and she was hindered from fixing them because she was attached to the dialysis machine. Desperately, she would look around to find the nurse who could help her put the needles back in her arm. Even though blood would pour from the opening on her fist, she felt no pain in her dream. When she woke up everything would be normal. Once while having a similar dream during dialysis, the dream materialized: Zehra woke up to find that her needles had come undone. Unlike in her dream, there was no blood. Just as in her dream, she did not feel pain.

After a while, Zehra began to feel kinship ties to the other dialysis patients—a "new family" as her fellow patients called themselves. The repetition of dialysis teaches the sufferers to recognize the same kind of suffering on the faces of their fellow patients. Beyond this, for dialysis patients, a new kinship is based on sharing a cyborg-like life and dreams that reveal suppressed feelings about this life. During their three-hour dialysis sessions, always scheduled for the same times every week, patients talk about family, pain, kidney thefts, the "organ mafia," the rich and their luck, school, flirts—in short, about everything one would talk about with close friends. Even though these patients have not known each other for long, they feel intimately connected. The family metaphor is reinforced through lack of choice: just as one cannot choose one's own parents, one cannot really choose one's companions in dialysis.

Zehra was not very talkative during her dialysis sessions. She wondered to herself if she was introverted because a childhood fever had left her with a limp, or perhaps she was simply too shy to ask questions. She preferred to sleep despite her nightmares. She wanted a dream state and silence, but the *dede* (a grandfather, or an Alevi saint), her next-bed neighbor, was a talkative older man, and it would be disrespectful to close one's eyes and pretend to be asleep while he was talking to her. So sometimes she would listen to him, and sometimes Zehra would try to sleep.

One day, during dialysis, Zehra turned her head toward the *dede* and saw something miraculous. He had changed. Now, his nose was just like hers, curved; it had changed its shape. His hands became like her hands, his thumbs became like hers, which were flat and narrow, and he had henna all over the inside of his palm. It occurs to her that he might be more than dialysis kin. Had she not realized before how similar they (the *dede* and herself) were, and how different she was from her own family? She felt related to the *dede* and not to her own parents. Even though she was not taking any drugs known for their hallucinatory effects, Zehra preferred to keep her insight about the *dede* a secret from the doctors. She thought she might be laughed at or put on more medication. This experience marked, for Zehra, the beginning of feelings that Virginia Woolf expressed when she described a moment: "when life sank down for a moment, the range of experience seemed limitless."[3]

Zehra saw other things. The doctors' skin and hair color would change as they approached her. At a distance, one doctor looked dark-skinned with dark eyes, but as he came closer his hair turned grey and his eyes became green. Another time Zehra saw a group of doctors who were all dressed the same: all were in black—an unusual color for the hospital. She was sure there was a change, but she was not sure of its cause. Was it the light or herself? If it was the light, it was merely an optical illusion. But if it was herself and if exterior objects were indeed changing to reveal their true appearance to her, then she had insight into the miracles of a world unseen by others. In this second case, she would not be able to talk to anyone about what she was seeing. Zehra began to wonder whether she was losing her sanity or whether the world was revealing its secrets. She felt a deep desire to know the truth. On several occasions, she had already felt the dark ether of death next to her bedside.

Then, one day, after being in the hospital for one-and-a-half years, Zehra fainted during dialysis and was taken to the emergency room. She had always found the hospital an unpleasant place, with most of the nurses and doctors overworked. She had rarely felt well treated. After her recovery, the fainting episode provided a reason to change dialysis units. Now she would go to a private hospital. To get there, she rode a private shuttle bus to and from the dialysis center three times a week. During each half-hour leg of the journey, Zehra took part in many conversations. Everyone on the shuttle was suffering from the same illness. Most of the conversations were on religious or spiritual issues. On the bus, the patients talked about death and the afterlife, how Sunnis, who made up the majority of Turkey's Muslim population, would go to Heaven and how the minority Alevis would suffer a purgatory.

At the age of twenty-seven, Zehra, a Muslim, questioned her religious and ethnic origins for the first time in her life. "I was not so sure what we were . . . so, I asked my mother. 'Are we Alevi?'" Her mother told her that they were. This distressed Zehra. If the Sunni women in the shuttle were right, then all of her Alevi practices were wrong. If the women in the shuttle were right, she would go to Hell like all the other Alevis. Of course, if the Alevis were right, then Sunnis would burn in Hell. But why, she asked herself, would so many people choose something so wrong? In Turkey, the Alevis made up about a quarter of the population, the Sunnis three quarters.

Life at home continued as usual, but Zehra felt her own life was coming to an end. Her illness kept her busy: she had to follow a strict diet and most of the time she suffered from low blood pressure. In the evenings, she was often too tired to do anything. She had been to dialysis and she had usually helped out in her uncle's small grocery store in their Istanbul neighborhood, a job she had taken after working in a series of sweatshops after high school. Usually, on Thursday evenings her family had people over. Family members sang gospels, read, and discussed verses from the prayer book *Berat Kitabi*. Once, Zehra was so tired that she could not join the prayers. This started an argument with her brother that soon got out of control. She told him that she was worried that as an Alevi she was on the wrong path; she told him that it was easy for her family to ignore this problem because none of them were dying, but that she would face the truth soon. What if she burned in Hell? Hearing her self-doubt, her brother got angry with her and yelled: "You are not one of us (*biz*)!"

That was it, this was what she suspected. She must never have been one of them; if she were she would be as healthy as they. She might be related to the *dede* who looked like her. Just a few days before her scheduled transplant surgery, she saw a dark shade approaching the left hand side of her bed. It was *Azrail*, the dark angel, who came to collect souls.

Zehra received her father's kidney late in 2001. I met Zehra shortly after her operation and spent two days with her in the hospital. Her frail body was weakened; she had a soft but tired look in her eyes. She shared with me what she hid from the others. She was not sure if the man who claimed to be her father was truly her father. "I will know if my body rejects the kidney (*bedenim böbregi atarsa*)," she told me. All the pre-surgical tests had gone smoothly; tissues and blood types matched. She thought he must have been her father, but, in her mind, her recent experiences had raised questions, not only about her spiritual affiliation as an Alevi but also about her parentage. If her body accepted the kidney, she reasoned, it would confirm her father's identity. And finally, she would know if *Azrail* would come back to collect her.

In this inner dialogue, we see a manifestation of Zehra's psychosis. Hers is a cultural psychosis common in Third World countries. Psychiatric terminology coins it "Nonaffective Acute Remitting Psychosis." In it, the subject suffers anxieties and panic attacks in which cultural and political images figure as persecutory or redemptive actors. Those who suffer from the disease share an experience of a particular kind of repression. In their cultures, authorities—whether political, economic, or religious—enforce incoherency. In the cacophony that results, traditions lose their functions, and citizens, lost and anxious, are plagued with self-doubt. In response, they seek refuge in alternate identities. So, in Zehra's case, being an Alevi began to feel intolerable. An imagined Sunni identity comforts her by making her part of the majority.

After the operation, and still uncertain of her father's real blood tie with her, Zehra was released from the hospital to rejoin her family. She would no longer leave home for dialysis; after her recovery, she again could work with her uncle in the market. But Zehra began to feel that people were looking at her from the television and were speaking to her from behind its screen. Her family was very kind to her now, her sisters too considerate, too nice, too quiet. When she watched the news, Zehra felt certain that the speaker was looking her in the eye, talking directly to her.

Zehra's new experience of the world, her schizophrenia, started with her uneasy partnership with the dialysis machine. Through the phases of her illness her symptoms evolved: her memories modified her sense of reality; she lost a sense of what was actually happening. She tried to bring order to her life. She found new ways to know herself and others. It was in this context that, during the dialysis, she had seen the old man next to her changing. This man, whom she called *dede,* represents more than an elder among the minority Alevi. A *dede* is someone who, like a shaman, conducts the ritual *cem* gatherings as part of Alevi communal life.[4] But a *dede* is a saint for material life. His knowledge does not come from divine sources but rather from experience and observations.[5]

When Zehra saw her ill neighbor as a *dede,* it made it possible for him to *almost* take her father's place in her life. She saw him as half-sacred and half-machine, half-her father and half-herself. Zehra was drawn to form a new kinship tie without a marriage bond, but through a technical connection with a machine. From Mary Douglas we learn that objects can only be classified as pure if a cosmology-maintaining line is drawn between the clean and the dangerous.[6] The *dede* is such a pure object. The changes Zehra sees in the *dede*'s face cause her some anguish, but they are soul-soothing illusions that enable her to identify with him. Through him, she can be part of the "*biz,*" the "us."

Among Alevis, the category *biz* signifies not only that their community differs from the Sunni population in Turkey, but the significance of this difference. During prayer and community meetings (*cems*), men and women gather to talk about politics, listen to elders' stories, interpret past events, and talk of daily life concerns. As they do so, they enact the power of *biz* in opposition to *ötekiler* (the others).[7] When Zehra's brother told her "you are not one of us, one of *biz,*" he was referring to her biological family and its historic genealogical roots in the whole Alevi community. This marginality was profoundly destabilizing for Zehra. She urgently needed to search for a truth and a sense of an after-life. She began to read popular religious and psychology books. But she was paralyzed by her liminal state, betwixt and between any stable notion of belonging.[8] In her anxiety, her identifications moved to the other patients in dialysis, the *dede* who was becoming more like her, and the doctors who seemed to be showing her a new truth as they changed under her gaze.

Where once the dialysis machine had sustained her, now other technological objects surveilled. Speakers from television screens spoke to her, guided her. With vigilance, she would read the world through a new world of signs, much as a person devoted to science or literature.

Oguz

> Men have called me mad; but the question is not yet settled, whether much that is glorious, whether all that is profound, does not spring from disease of thought—from moods of mind exalted at the expense of the general intellect. They who dream by day are cognizant of many things, which escape those who dream only by night. In their gray visions they obtain glimpses of eternity, and thrill, in awaking, to find that they have been upon the verge of the great secret. In snatches, they learn something of the wisdom, which is of good, and more of mere knowledge, which is of evil. They penetrate, however rudderless or compassless, into the vast ocean of the "light ineffable."
>
> —Edgar Allan Poe, "Eleonora," *The Murders in the Rue Morgue*

Twenty-six-year-old Oguz's health problems began sometime in 1996. While walking in a strange neighborhood, he was beaten up by a group of young men. Oguz said he did not know who they were or why they would attack him. He went home covered in blood; he could not touch his chest. Internal bleeding sent him to a university hospital and, after over a month of tests, he was diagnosed with a rare illness caused by lung damage. Oguz's kidneys began to fail, and he was sent to the hospital's dialysis unit. He remained there for three years.

Oguz found the dialysis unit an unfriendly place. "One accepts being a part of an experiment once one is in the hospital," he told me. Oguz describes his examinations as humiliating, objectifying. At least ten different doctors would come by, together with teams of assistants. As they touched him, examined him, discussed his files, took blood samples, he felt that they spoke of him as an alien being. Apart from feeling like a research object, dialysis was uncomfortable. He describes his feelings with heavily loaded phrases. "*Öldüm öldüm dirildim. Normal insanlıktan çıkmış oldum.*" The literal translation of these words is: "I died, I died, and then was resurrected. I exited the human condition." After dialysis, Oguz would repeatedly vomit on the minibus he took to get home. Once there, he would be so weak that he could hardly stand. His days became unbearable and he decided to switch to a private dialysis center.

In the new center, his situation improved. He still felt less than human, but now he found a room where he could socialize with the people who worked at the facility. The doctors there were friendly and reassuring. He would have tea with the owners of the center, a retired high school teacher and his wife, and they would make social conversation or discuss matters about which he could not speak easily with his family. The center became a new home. He felt welcomed like a son. His vomiting ceased. After dialysis, he went home, tired and relaxed.

Yet being on dialysis—three hours, four times a week—made Oguz feel "half-robot and half-human."

His robot self needed to be "charged" by a connection to a machine. When Oguz left the hospital, the small plastic apparatus on his arm that connected him to the dialysis machine was left open even when he was unplugged from the machine. The opening made him feel as though he was leaking, without boundaries, open to domination. Oguz knew the importance of his machine connection: if he missed an appointment, he would go into a coma. If he had been older, he would probably have died.

In dialysis he learned what suffering was. "*Bilmek degil, yasamak önemli,*" he says. "It is not what one knew, but what one experienced, lived through, that made things real." He met the secrets of Poe's "gray visions." He began to see his body as an electronic thing to be tended. Once, on a beautiful summer day in Istanbul, he went swimming after a dialysis treatment. A few days after his swim, large wounds appeared on his back. He reasoned to himself that immersing a robot in water would have broken its mechanism in just this way. The experience confirmed for him that he was a robot, just like Arnold Schwarzenegger in *The Terminator.* He felt more a machine that would break in water than a human who would not.

Oguz's body was deteriorating as it became alien to him. Dialysis fluids changed his skin's immunity and resistance. Infused with dialysis fluid, his wounds bled a lot; cuts took a long time to heal. Connection to the dialysis machine eroded the boundaries of the ego and the resilience of the body.

Oguz was not allowed to drink more than three liters of water between dialysis sessions. But he would often drink nearly double that amount, swelling his body. Then, after he had lost the water in dialysis, Oguz felt emptiness inside him, such as that a melancholic might feel, a void that could only be mourned or filled back in.[9] Oguz tried to fill the void by drinking with friends. He tried to feel like a man of his age, someone healthy and normal, even though his doctors forbade alcohol. In the small secret room behind the local *bakkal,* he drank with his friends until the morning hours.

Oguz suspects that his doctors knew about his self-destructive behavior, although they never discussed it with him. The doctors were working with his family to find a potential donor for a kidney transplant. His father, a positive match, offered to be his donor. Unsure of how the transplant would change his life, Oguz said he did not want it. But the doctors were doing more than suggesting the transplant—they had moved to telling him how to prepare for it with diet and medication. In the end, Oguz received a kidney from his father. He told me: "Had I known that it would be this way, I would not have accepted it." Father and son had not been close; now Oguz felt indebted to his father.

Not until after the operation did his doctor pull him aside to tell him he should not have sex or marry. Oguz decided to marry his girlfriend anyway. When he announced that he was engaged, many patients got angry with him, telling him it was not his right to destroy the life of a young girl by risking having a baby or by preventing her from having one. "What if the children suffer from the same disease? What if you die?" they asked him. What if he died? He did not want to think about this any longer, because the question of death occupied his mind all the time, so much so that one day, he attempted suicide.

> A year ago, I made a suicide attempt and I was taken to the hospital. There I visited a psychiatrist, but I did not tell her anything. In dialysis, a psychologist approached me, but I did not say anything. Now I am talking to you. After the dialysis, you feel empty. When you talk it echoes, so you are like an empty thing. . . . So I got something from the pharmacist over the counter. I came home, had a little argument with my parents, and

then went to my room. I smoked two cigarettes. I swallowed all the medication. I waited. . . . In the meantime, one of my parents came and we had another short argument. . . . I felt my head turning, so I left home only with my pants on, no shirt, no shoes. I went to the school. In the meantime my mom called all my friends, sent them to look for me. It took them one and a half hours to find me. I had my mobile phone but I didn't answer. If they had been an hour later than they were, we— the kidney and I—would have been gone. Then, I started therapy. There the doctor asked me odd questions I did not answer. He asked me if I was sexually abused when I was a child. I said no! I said maybe *he* was abused, not me! I got so upset with his attitude.

Oguz did not remember why he wanted to commit suicide, but he tells me that with the transplant he felt trapped. He could not eat or drink, or go out to watch a soccer match. Each time he did, the kidney failed, and he had to go back to the hospital for a few days of dialysis. With his arm opened, and then a new kidney placed in his body, he no longer felt himself to be a true biological creature. During an argument, a close friend got upset and yelled at him: "Do not make me angry or I'll punch you and then you are done with. . . . You are half-a-something anyway . . . (*yarım kalmış bir şey*)!"

Oguz could not forget this incident. Others saw him as "half," unfit as a companion. Oguz now thought that his life on dialysis had been a better life. On dialysis, he could drink and socialize. His body would swell from the alcohol with which he abused it, but then he could be cleansed by the machine. The machine made it easier to live a "normal" life compared to the frail kidney he had received from his father. This kidney not only needed care, but after his transplant, the doctors told him that it would last at most for nine years.[10] It was the machine not the new kidney that was a solution for "forever." In time, when his body would reject his father's kidney, he would go back to the machine again. Then why not go back now? Why wait for nine years?

Oguz changed when he began living with his father's kidney. He used to love the smell of soap. He would wash his hands the minute he arrived home. He would use up four or five bars of soap every week. He would play with the soap bars. Now as he puts it, "All these habits changed." He would no longer wash his hands. He said: "I have not washed my hands for two days. Since the transplant I can no longer wash my hands. . . . If only I knew why, if I knew why, maybe I could start playing with soap again."

A few months after the transplant, Oguz cut off the Gautier-style beard he had made popular in his neighborhood. At one time he had shaved his beard into seven different shapes, a different style for every day of the week. Now, he would not look in the mirror or iron his trousers. He hardly went out. Since childhood, his father had been the person with whom he least wanted to identify. Having his father's kidney made it impossible for him to avoid identifying with his father. And with this identification came an anxiety he could not overcome.

The Influencing Machine

In 1919, the psychoanalyst Victor Tausk wrote about a group of schizophrenics with a particular paranoid delusion.[11] They believed that their problems were caused by an "influencing machine" operated by alien forces. The patients saw the machine as feeding on the emotions and "souls" of human beings unconscious of their true state. Indeed, for these patients, knowing about the machine, that is "seeing" the real, could be fatal because knowledge of the machine's existence revealed the givens of everyday reality to be fabrications. For those taken up in this delusion, human subjects are, bit by bit, becoming part of or merged with the machines.

Tausk notes that, in the main, his patients became delusional at the time of a surgical intervention. At that time, the sense of invasion and loss of control diffused into the metaphor of machine invasion and control. And to this day the use of machine metaphors marks persistent fears of invasion, possession, and authoritarian control.

For Zehra and Oguz, with their transplants, first their bodies opened and then their selves. Weakened, left without a sense of outer or inner limits, they feel vulnerable, at the verge of extinction, and try to find something to turn to. Zehra discovers a new relative in the *dede*. She moves toward new allegiances, beginning to identify with the majority Sunni population. Oguz takes on his father's habits, for example, not washing his hands or shaving. He tries to save himself with a new "robot identity," fashioned in the image of Arnold Schwarzenegger's action heroes. He wants to feel invulnerable, strong enough to consume alcohol strong enough to resist turning into his donor. Both Zehra and Oguz try to become stronger by internalizing the most powerful objects available to them in their culture: a *dede,* a Sunni identity, a *Terminator* self. Such identities give a feeling of becoming whole again. They serve as objects-of-rescue, suggesting the existence of another order underlying the surface of events, an order that is not religious but grander than the everyday. For Zehra and Oguz, the world is neither completely theirs nor truly God's. In our high-tech imaginary, the influencing machine intimates or replaces the divine.

Aslihan Sanal is a cultural anthropologist who studies the political economy of biomedical and biotechnological practices in the Middle East and Europe. She lives and works in Paris.

VIDEO POKER

Natasha Schüll

Patsy, a green-eyed brunette in her mid-forties, began gambling soon after she moved to Las Vegas from California in the 1980s with her husband, a military officer who had been stationed at Nellis Army Base.[1] Video poker machines had been introduced to the local gambling market in the late 1970s, and she discovered them on her trips to the grocery store. "My husband would give me money for food and milk, but I'd get stuck at the machines on the way in, and it would be gone in twenty minutes. . . . I would be gone, too—I'd just zone into the screen and disappear."

Ten years later, Patsy's gambling had progressed to a point where she played video poker machines before work, at lunchtime, on all her breaks, after work, and all weekend long. "My life revolved around the machines," she recalled as we talked outside the Gamblers' Anonymous meeting where we had met, "even the way I ate." Patsy dined with her husband and daughter only when the three met in casinos; she would eat rapidly, then excuse herself to go to the bathroom so that she could gamble. Most often she gambled alone and afterward slept in her van in the parking lot: "I would dream of the machines. I would be punching numbers all night." Eating alone, sleeping alone, Patsy achieved a sort of libidinal autonomy. Her time, her social exchanges, her bodily functions, and even her dreams were organized around gambling. "When I wasn't playing," she told me, "my whole being was directed to getting back into that zone. It was a machine life."

The French sociologist Roger Caillois, author of *Man, Play, and Games,* believed that games were "cultural clues."[2] One could diagnose a particular civilization, he wrote in 1961, by examining its games. Erving Goffman undertook such a diagnosis in a 1969 study based in Las Vegas, describing gambling activities as the occasion for "character contests" that functioned to maintain social cohesion.[3] In 1974, Clifford Geertz similarly interpreted Balinese cockfight gambling as a "tournament of prestige" that simulated the social matrix, rendering selfhood and social mechanics visible to individual participants.[4]

Caillois, Goffman, and Geertz all referred to slot machine gambling in the course of their analyses: for Caillois, it was an absurd, compulsive form of play in which one could only lose; for Goffman, it was a way for a person lacking social connections "to demonstrate to the other machines that he has socially approved qualities of character"[5]; for Geertz, the gambling machine was a "stupid mechanical crank" offering play that could be of interest only to women, children, the poor, and the despised, and therefore not a properly "sociological entity."[6] The fact machine play has come to drive the American gambling economy prompts me to reconsider these appraisals. Perhaps there is something important taking place that Geertz and his colleagues were not in a position to see.

Although classic green-felt table games or "live games" dominated casino floors as recently as twenty years ago, today profits come primarily from machines, which in residential areas of Las Vegas earn as much as 89 percent of casino revenue—a figure that grows higher still when one considers the machines in gas stations, pharmacies, and supermarkets across the city.[7] At local Gamblers' Anonymous meetings, the vast majority of the participants play video poker exclusively. Robert Hunter, a charismatic Las Vegas psychologist who has carved out a therapeutic niche in the treatment of gambling addiction, has referred to video poker machines as "electronic morphine," "the distilled essence of gambling," and most famously, "the crack cocaine of gambling."[8]

While all contemporary slot machines offer a choice of how many credits to bet on each spin, video poker goes a step further by allowing players to decide which cards—of those they are "dealt" by the machine—they wish to hold or discard in order to make winning hands.[9] As Hunter understands it, the technological interface of video poker so completely concentrates players' attention on a series of specific game events that anything troubling about one's life situation—physically, emotionally, or socially—is blotted out.[10] Unlike Geertz's cockfight, machine gambling is a nonrelational activity in which gamblers enter a dissociative state—a "zone," as they call it—in which a sense of time, space, the value of money, social relations, and even a sense of the body dissolves. "The consistency of the experience that's described by my patients," Hunter told me, "is that of numbness or escape: video players don't talk about excitement—they talk about climbing into the screen and getting lost."

The literature of gambling behavior has hypothesized a qualitative split between *action* gambling and *escape* gambling. Historically, this split has been analyzed in terms of gender dichotomies: men are categorized as action gamblers who prefer live games and women as escape gamblers who prefer machines; men play for sociability, competition, and ego enhancement while women play for isolation and anonymity; men seek thrill, excitement, and sensation while women seek to dull their feelings, escape distressing problems, and relieve themselves of interpersonal interaction.[11] Whether or not these distinctions held true in the past, these days video gambling in Las Vegas is increasingly seductive to men and women alike. At an average Gamblers' Anonymous meeting 97 percent of those in attendance play only machines, and the gender split is even. A general turn to escape gambling seems to be taking place.

The predicament of compulsive gamblers in Las Vegas, for whom a desire to lose themselves in machines trumps a desire to be selves in the world, can be understood not only as a symptom of gender, genetic predisposition, trauma, or individual life circumstances, but as an index of larger tensions surrounding ideals of individual autonomy in free-market society. It is not my intention to aestheticize or romanticize the plight of compulsive gamblers, but rather, as Jackson Lears puts it in his book on gambling in America, to approach their experience "as a port of entry into a broader territory."[12] I look to the dramatic turn away from social forms of gambling, played at tables, to asocial forms of gambling, played alone at video terminals, for clues to the discontents of today's self. The turn to machine gambling mirrors the spread of consumer technologies like video games, personal computers, and the Internet—technologies, that have given rise to new and intensified forms of human-machine exchange, along with new sets of anxieties for selves, and new forms of self-escape.[13] Here, I parse the escape that Patsy calls "machine life" into its different forms: escape from bodily existence; escape from social exchange; escape from monetary value; escape from chronological time. What about being a self today do compulsive gamblers wish to escape, and how do slot machines become places of refuge?

The Self as an Enterprise

Since the late 1970s there has come into cultural circulation a model of the self "as a kind of enterprise, seeking to enhance and capitalize on existence itself through calculated acts and investments."[14] This model lends to

selfhood a calculative vocabulary of "incomes, alloca-
tions, costs, savings, even profits" by which individuals
can engage in a kind of maximizing self-accounting.[15]
Nikolas Rose elaborates: "Numbers, and the techniques
of calculation in terms of numbers, have a role in sub-
jectification—they turn the individual into a calculating
self endowed with a range of ways of thinking about,
calculating about, predicting and judging their own activities and those of others."[16]

The movement of financial management technologies into the realm of selfhood does not indicate a new
"corruption" of the private by the public. On the contrary, the self has long borrowed its metaphors from
enterprise.[17] In today's free-market societies, self-
enterprise has become a dominant model of selfhood
that Pat O'Malley calls "privatized actuarialism."[18]

Choice is key to this reframing of the self; it defines both the field in which actuarial selves operate
and their behavioral imperative. Rose, following Anthony Giddens, has put it thus: "Modern individuals
are not merely 'free to choose', but *obliged to be free,* to
understand and enact their lives in terms of choice."[19]
For psychologist Barry Schwartz, upper- and middle-
class citizens tend to associate choice with freedom, action, and control while working-class citizens tend to
associate choice with fear, doubt, and difficulty. Nevertheless, today in America "the equation of freedom with
choice is the one most loudly broadcast."[20]

In his bestselling book, *The Paradox of Choice:
Why More Is Less,* Schwartz argues that added options
do not necessarily make us better off as a society, as so
many economists, policymakers, social scientists, and
citizens assume. Too much choice, he contends, can
overload, debilitate, and tyrannize.[21] Sifting through an
"oppressive abundance" of choice takes up time and
energy, causes self-doubt and anxiety, and can multi-
ply occasions for disappointment, regret, and guilt. We
are left "feeling barely able to manage" our lives.[22]

Others take his point a step further, arguing that selves burdened by demands for autonomy and responsibility are liable to turn to self-annihilating, self-defeating forms of escape.[23] One might go even further, for it is not simply the demand for autonomy that burdens people but the fact that this demand is quite difficult to realize in a society that protects the welfare of its markets above the welfare of its citizens. Individuals are exhorted to become maximizing virtuosi of the self, but they are not given the material support that would enable them to do so. Even where such support is available, the "ideal self" is by definition never complete, never developed enough. The project is ongoing, driven by a sense of incompletion.

If lives are caught between the task of optimization and a sense of insufficiency, it is not surprising that there arises a desire for relief. Nor is it surprising that the kind of relief sought is one that distorts and denies the premises of the actuarial self—not necessarily by rejecting these premises but often by taking them to an extreme.

Self-Liquidation

Physical boundedness—the material basis for self-hood—dissipates in the zone of machine play. Isabella, a real estate agent and single mother, likens her absorption in gambling machines to the way in which the bodies of characters on a science fiction television program are sucked into the screen:

> On TV they express it by *pulling*—the bodies actually disappear into the screen and go through the games of the computer. That's what gambling on the machines correlates to: for the time that I was there I wasn't present—I was gone. My body was there, outside the machine, but at the same time I was inside the machine, in the king and

queen turning over, almost hypnotized into *being* that machine.

Like Isabella, an insurance agent named Josie speaks of exiting her body and entering the machine through a kind of pulling: "You go into the screen, it just pulls you in. You're over there in the machine, like you're walking around inside it, going around in the cards."

Absorbed in play, gamblers are apt not to attend to the functioning of their bodies. One regularly found, after sessions of play, that she had unwittingly vomited on herself, and once, that she had wet her pants. Another wore double layered dark blue wool pants so that she could urinate without leaving her machine. Human and machine seem to merge, as an electronics technician named Randall describes: "I get to the point where I no longer realize that my hand is touching the machine, I don't feel it there. I feel connected to the machine when I play, like it's an extension of me, as if physically you couldn't separate me from the machine." Josie remarks: "It's like playing against yourself—you are the machine; the machine is you."

Machines allow gamblers to forge an autonomous, insulated zone of play. Julie, a fundraiser, says "People break the flow and I can't stand it—I have to get up and go to a machine, where nobody holds me back, where there's no interference to stop me. I can have my free rein, go all the way with no obstacles." A blackjack dealer named Sharon says:

> In live games you have to take other people into account—other minds making decisions. . . . Like when you're competing for a promotion, you're dealing with other people who decide which one is the best—you can't get into their minds, you can't push their buttons, you can't do anything about it. Just sit back and hope and wait. When

you're on a machine you don't compete against other people.

The "live" world, as Sharon describes it, is a kind of relentless character contest demanding that she "take other people into account" yet providing no clear feedback on which she might base her calculations or hedge her bets. The need to take others into account contrasts with life on the machine, which protects her from the nebulous and risky calculative matrix of social interaction. "At the machines I feel safe, unlike being with a person. [On the machines,] if I lose, that's the end of the relationship. Then it starts again, fresh."

"Nobody really talks to each other when they're playing video poker machines," Patsy tells me, "so just about anywhere you sit you're isolated." Some gamblers select machines in corners or at the end of a row; others place coin cups upside-down on adjacent machines to prevent others from sitting down. "I want to hang a DO NOT DISTURB sign on my back," says Sharon, who has learned to buy a liter of coke and two packs of cigarettes before sitting at the machines so that cocktail waitresses will not interrupt her.

"If you work with people every day," says Josie, "the last thing you want to do is talk to another person when you're free. . . . You want to take a vacation from people. With the machine there's no person that can talk back, no human contact or involvement or communication, just a little square box, a screen." She links the escape of machine play to the taxing social exchange that her job demands:

> All day long I have to help people with their finances and their scholarships, help them be responsible. I'm selling insurance, selling investments, I'm taking their money—and I've got to put myself in a position where they will believe what I'm selling is true. After work, I have to go to the machines.

In the digitized safety of machine interaction, Josie seeks relief from the actuarial logic of her vocation and the calculative interpersonal engagements it entails. "Nobody talked to me, nobody asked me any questions, nobody wanted any bigger decision than if I wanted to keep the king or the ace. At the machines, I was safe and away."

Josie suggests here that the turn to asocial forms of gambling may have something to do with the growth of service sector labor. If, as sociologist Daniel Bell suggests, postindustrial society is an economy driven by exchanges between people rather than between people and machines, then perhaps it makes sense that people are dedicating increasing amounts of their time away from work to machines.[24] The machine is a respite. In Las Vegas, a city that sociologist Mike Davis has called the "Detroit of the postindustrial economy," the hypersociality of labor forms is accompanied by forms of escape that seem to reproduce asociality as a norm.[25]

In *The Managed Heart*, sociologist Arlie Hochschild argues that a shift from assembly-line factory work to service sector work has been accompanied by a shift from physical labor to what she calls "emotional labor," or labor in which "the emotional style of offering the service is part of the service itself."[26] Although physical labor requires detaching oneself from one's own body, emotional labor requires detaching oneself from one's own feelings; workers are prone to alienation or estrangement from their emotions, which wear thin as they are processed and managed in the marketplace.[27] Hochschild's analysis resonates in gamblers' narratives. "There are so many people dependent on me, I have no sense of self," says Lola, a buffet waitress and mother of four: "When I'm at the machines, all my obligations fall away and I can fade away."

Money and Time

Patsy recalls her job as a welfare officer at the State of Nevada's food stamp office: "All day long I'd hear sad stories of no food, unwanted pregnancy, violence. But it all slid right off me because I was so wrapped up in those machines. I was like a robot—*next, snap, what's your zip code, you don't belong here*—I wasn't human." "The machines were like heaven," she remembers, "because I didn't have to talk to them, just feed them money."

Money is a key resource of the actuarial self. Its value shifts through the digitized repeat wagering of machine play. Julie told me: "You put a twenty dollar bill in the machine and it's no longer a twenty dollar bill, it has no value in that sense. It's like a token, it excludes money value completely." Josie elaborates: "Money was almost like a God, I had to have it. But with the gambling, money had no value, no significance, it was just this thing—just get me in the zone, that's all . . . You lose value, until there's no value at all. Except the zone—the zone is your God."

In the economy of the zone, winning money is not the aim. "You're not playing for money," says Julie, "you're playing for credit—credit so you can sit there longer, which is the goal. It's not about winning; it's about continuing to play." Sharon would rather "spend two hours losing a jackpot" than cash it out, as this would mean exiting the zone to wait for the machine to drop her winnings, or, in the event that its hopper is low, for attendants to come pay her off. Winning—too much, too soon, or too often—interrupts the tempo of play, disturbing the harmonious regularity of the zone. Julie explains: "If it's a moderate day—*win, lose, win, lose*—you keep the same pace. But if you win big, it can prevent you from staying in the zone." Continuance of play depends on money's conversion from a means of acquisition to a means of suspension within a closed, autonomous circuit.

"Gambling changed my relationship to money" said Isabella. "Money became the means to gamble, that's all it was to me. I'd conserve gas so I'd have the money to gamble, and instead of going to the grocery store regularly, I'd wait to go to Wal-Mart and do it all at one time—that way I wouldn't have to waste the gas to go more than once. I *economized*." Economizing—the responsible accounting behavior of the risk-managing self—is harnessed to radically different ends by compulsive gamblers.

In between play sessions, Patsy engaged in compulsive calculative rituals around money:

> For me, getting the money together was part of the process. I'd go to the bank and get $1,000, $400, whatever amount. I had a weird thing where I could never just take out $20, or just spend $43—I had to spend in $100s. And other weird things too. . . . Like if I won, I could spend back to $500 but I would never keep $600; it would be okay to put back $800, but I had to keep another certain amount—there were lots of little rules.

After gambling, Patsy would sit and count her money, "over and over again, in my car, at stop lights in the dark, in my lap, hundreds of dollars—what was the use?" Her excessive attention to money, she suggests, was beyond utility—a "weird thing" that served no clear purpose. Money became fetish-like, unhinged from conventional exchange value. "I spent a lot of time thinking about money, touching money, calling the bank to keep track of my money, to know the time frame of when checks cleared, counting it and counting it . . . but in fact, I wasn't actually *counting* at all." The year after Patsy stopped gambling she did her taxes and found that over a six-month period of gambling, when she had not been able to "count" according to conventional value standards, her losses exceeded $10,000.

"It stopped mattering how much I won, or lost," Sharon remembers of her ten years of gambling. "I couldn't set limits. The last time I gambled I was playing with a hundred dollars and I lost it all. I parlayed it into $200, then got down to one dollar, and with that I hit another hundred dollars. I didn't leave until I spent the whole hundred dollars again." Julie breathlessly describes the sequential refinancing of a typical play session:

> I got four aces four times, that's $200 a shot, 800 credits each time, that means I could have cashed out $800 total. But each time I hit, I'd play it down to 200 credits from 800 credits and I'd say, Well, I'll just hit the aces again and then I'll leave. Then I'd get four of a kind and have like 437 credits and I'd say I'll just go to 400 and leave, and then at 400 I'd just push the button again and drop below 400, and I'd say, Well now I'm down past 400, I'll just get back up to 400 and then I'll cash out. And then I'd find myself closer to 300 and I'd say, Once I get down to 300 I'll go. And then when I go below that I'd say, Well, I might as well keep going, I've already blown what I was gonna blow—I might as well try to get the aces again, and it would continue.

Dizzying swings in fortune distort all sense of value, such that gains and losses are difficult to track and blur into each other.

In his book on gambling, Lears asks: "In a society such as ours, where responsibility and choice are exalted, where capital accumulation is a duty and cash a sacred cow, what could be more subversive than the readiness to reduce money to mere counters in a game?"[28] He suggests that gamblers pose a challenge to the idea of productive citizenship: "The idea that loss is not only inescapable but perhaps even liberating does not sit well with our success mythology, which assumes at least implicitly that 'winning is the only thing.'"[29]

Compulsive gamblers complicate Lears's analysis, suggesting that something more is afoot that a simple challenge, subversion, or liberating renunciation of money. In their own way, gamblers act within social values about money. Patsy, the welfare officer, told me: "I always had income coming in, every week it was something—a $600 paycheck, $500 child support, my husband's retirement checks. We always had like three credit cards so if I had a bad spell I'd just put it on the cards." The resources of a conventional calculative lifestyle supported Patsy's gambling, and vice versa: "One time I had maxed out the three cards, but then I hit a jackpot and paid them all off." Her compulsive gambling does not exactly oppose the logic of the risk-managing, maximizing self—if anything, it intensifies (or "maxes out") that logic. In this sense, Patsy's gambling could be said to convert the calculative ideals of free-market society into a mode of behavior that departs from dominant models of productive citizenship.

Time is another resource of calculative selfhood that gamblers manage to convert into a means of escape through their machine play. In the play zone, the measured pace of "clock time," as gamblers call it, "stops mattering," "sits still," is "gone" or "lost"; time seems to condense or expand according to the particular rhythm they establish. Compulsive gamblers, for whom play sessions of seventeen hours or whole weekends are not unusual, talk about time as they do about money: one can *salvage* from time, *extract* from it, *liquidate* it, *squander* it. Time becomes a kind of credit whose value is refinanced along the lines of play—a manipulable playing currency rather than a determinative order. Randall comments that machine play makes him feel he is "bending time" such that he goes "into a different time frame, like in slow motion. It's a whole other time zone."

Just as money value is converted into zone credit, clock time is converted into zone time. Julie describes this:

When the time comes to leave and the things I escaped from start crowding back into my brain, I find myself rationalizing, *Well, I don't really have to go today . . .* and I ask an attendant to hold my machine while I run to the payphone to buy myself more time—*Oh, I'm sick,* or, *I can't come today*—and then back to continue, and now there's three more hours. And when those three hours are up, I think, *I have to save money for the phone calls I'll have to make to cancel all the appointments I am going to miss. . . .* I'm thinking of how to arrange things so that I can stay there, how to economize.

The economizing of zone time, like that of zone credit, extends beyond the zone itself: "Time in general, not just while I was playing," says Sharon, "becomes very distorted. I feel like I can manipulate it very easily, salvage much more than I can from a small unit of it—go grocery shopping on the way to the casino, and while I'm there make a doctor's appointment on the cellular phone, and then on the way home. . . . Everything I do is relative to gambling time."

"I'd be later and later and later to work," Patsy recalls:

At break time I'd ask my supervisor, *Do you mind if I go to the bank?*—and I'd be out the door. I was just *wound.* My sense of time was totally out the door. I'd win a royal and I'd be ticked off because I'd have to wait for them to come pay me off. The other workers would look at the clock when I came back and I would think, *What are you looking at the clock for? Mind your own business.*

As Patsy describes it, her sense of time—and she herself—are "out the door." At every chance she attempts to escape clock time, such that she becomes her own kind of time: she is "wound" like a clock; she is "ticked off" as time ticks by during her wait for a jackpot pay-off; when she returns to work, resentful coworkers look pointedly at the clock.

Machine Life

> The mechanism to which the participants in a game of chance entrust themselves seizes them body and soul, so that even in their private sphere . . . they are capable only of a reflex action. . . . They live their lives as automatons . . . who have completely liquidated their memories.
> —Walter Benjamin, "On Some Motifs in Baudelaire"

> I hear most people say that the definition of gambling is pure chance, where you don't know the outcome. But I do know: either I'm going to win, or I'm going to lose. The interaction with the machine is clean cut, the parameters are clearly defined—all I have to do is pick *yes* or *no*. I decide which cards to keep, which to discard, case closed. I don't care if it *takes* coins, or *pays out* coins: the contract is that when I put a new coin in, get five new cards, and press those buttons, I am allowed to *continue.*
>
> So it isn't really a gamble at all—in fact, it's one of the few places I'm certain about anything. If I had ever believed that it was about chance, about variables that could make anything go in a given way at any time that I couldn't control, then I would've been scared to death to gamble. If you can't rely on the machine, you might as well be in the human world where you have no predictability either.
> —Sharon

If all gambling entails the "certain rapid resolution of an uncertain outcome," as Erving Goffman phrased it, then digital gambling further truncates the uncertainty, immediately resolving the event of the bet with the press of a button.[30] Sharon remarks: "I was addicted to making decisions in an unmessy way, to engaging in something where *I knew what the outcome would be*." Unlike Geertz's cockfight, the gambling machine is not a conduit of chance that allows selfhood to emerge in a context of social and economic uncertainty, but rather, a dependable mechanism for securing a zone of certainty in an uncertain human world. While the human world is associated with dependence, temporality, unpredictability, mess, and dangerous remainders, the machine is linked to dissociation, atemporality, predictability, closure, and depletion—a deadly, circular sort of perfection.

"The video game holds out two promises," wrote Sherry Turkle in 1984; "the first is a touch of infinity— the promise of a game that never stops. . . . [The second] is the promise of perfection.[31] She suggested that the appeal of video games lay in their consistency of response and "either/or" structure. Machine life, as Patsy calls it, shrinks choice to a limited universe of rules. All you have to do is "pick yes or no," Sharon tells us. "Awake, my whole day was structured around getting out of the house to go gamble. At night, I would dream about the machine, I'd see it, the cards flipping, the whole screen. I'd be playing, making decisions about which cards to keep and which to throw away." The digital game interface orders gamblers' waking lives and dream lives with its unending flow of mini-decisions.

What relation does this flow of mini-decisions bear to the ever-proliferating choices that actuarial selves face in contemporary free-market society? Compulsive machine gamblers speak of choice in ambivalent terms: it is emancipatory and entrapping, annihilatory and capacitating, reassuring and demonic. Lola, the buffet waitress, talks of "resting in the machine," then later in her narrative describes video poker's relentless stream of card-choosing as commanding—the activity does not only *hold,* but *hooks* and *captures* her attention. Julie identifies the compulsory nature of the choices she faces while gambling: "You *have no choice* but to concentrate on the screen; you simply cannot think about anything except which cards you are going to choose to keep and which you're going to choose to discard." Video poker compulsion unfolds in and through choice—precisely where a free-market logic posits the triumph of the actuarial self.

Randall tells me that he plays video poker because he wishes to be "in control"; moments later, without a sense of contradiction, he confides that he wishes he was "a robot," free of self-directive capacities. Video poker grants both of Randall's wishes. Although the game multiplies choices, they are reformatted as a self-dissolving flow of repetitive motion—so digitally intensified that one makes them without "choosing" as such.

In the zone of machine play that compulsive gamblers describe, the conventional means of self-enterprise—money, time, sociality, and bodily existence—are disengaged from the agenda of the maximizing self, and put in the service of self-liquidation. This self-liquidation participates in the logic of what Sigmund Freud called "repetition compulsion" or "death drive," by which he meant an organism's ongoing reflex to extinguish the perturbations, tensions and uncertainties of existence in order to return to a state of rest.[32] The detours or "circuitous paths" that each organism takes in moving toward this state of rest are what shape its life struggle, and in this sense, the death drive is vital. Compulsive gamblers embrace the gambling machine as a mechanism for short-circuiting life and its struggle.

Gambling machines, designed to accelerate "player extinction," distill the economy of the death drive such that it becomes an end in itself. Eventually, Sharon no longer even looks at the cards she has been dealt: "You reach an extreme point where you don't even delude yourself that you're in control of anything but strapping yourself into a machine and staying there until you lose. All that stuff that may draw you in the beginning—the choice, the decisions, the skill—is stripped away, and you accept the certainty of chance: the proof is the zero at the end."

Citing social phenomena as far-ranging as video games and religious fundamentalism, some contemporary psychoanalytic scholars suggest that the death drive has become a harbinger of an end to social ties. My aim here has not been to position compulsive gamblers as exemplars of a sweeping social diagnosis such as this, but rather, to examine the specifics of their experience for clues that might tell us more about the forms of self-loss that people seek today, and the forms of selfhood they seek to relieve.

Postscript

Until 2003, former "drug czar" William Bennett, Director of the Office of National Drug Control Policy, was best known for his philosophy of personal accountability. He elaborated this philosophy in *The Book of Virtues,* including chapters on self-discipline and responsibility.[33] A widely publicized scandal in 2003 revealed his substantial gambling habit; video poker was his game of choice. "I've been a machine person," he told reporters: "When I go to the tables, people talk—and they want to talk about politics. I don't want that. I do this for three hours to relax."[34] In fact, casino records show that Bennett often played for two or three days at a time, preferring the 500-a-pull slot areas in Atlantic City and Las Vegas casinos. Although he lost $8 million over a decade of gambling, he claimed that his habit was not technically an addiction because "I don't play the milk money."

What public figure better captures the uneasy embrace of self-sovereignty by citizens of free market society than Bennett—simultaneously a beacon of individual autonomy and of addiction?

Natasha Schüll is Assistant Professor in the Program in Science, Technology, and Society at MIT. She studies consumer technologies, the experience of addiction, and the role of neuroscience in society.

Notes

Inner History | *Sherry Turkle*

1. Several disciplinary traditions, including anthropology, sociology, and social psychology, use ethnographic or methods in which the investigator joins a social/cultural milieu as an observer or participant-observer and systematically reports his or her experience. In this volume, all informant and patient data protect confidentiality.

2. Clifford Geertz, "Thick Description: Toward an Interpretive Theory of Culture," in *The Interpretation of Cultures* (New York: Basic Books, 2000 [1973]), 21.

3. Said Geertz, "We are seeking, in the widened sense of the term in which it encompasses very much more than talk, to converse with them, a matter a great deal more difficult, and not only with strangers, than is commonly recognized." Ibid., 13.

4. Ibid., 20. At the same time that Geertz was mapping out an anthropology of interpretation, the psychoanalyst Jacques Lacan was trying to move his field away from medicine and clearly redefine it as a discipline of self-reflection. Controversially, he characterized this position as a "return to Freud." For Lacan, a "cure" may follow from the analytic encounter, but if it comes, it is *par surcroit,* as a kind of bonus or secondary gain. See Jacques Lacan, *Ecrits: A Selection,* trans. Alan Sheridan (New York: W. W. Norton, 1977); and Sherry Turkle, *Psychoanalytic Politics: Jacques Lacan and Freud's French Revolution* (Guilford, Conn.: Guilford Press, 1992 [1978]).

5. Anne Lamott, *Some Instructions on Writing and Life* (New York: Anchor, 1995), xxvii.

6. Geertz, "Thick Description," 17.

7. See also D. W. Winnicott's work on transitional objects, poised between a child's sense of self and sense of the outside world. See D. W. Winnicott, "Transitional Objects and Transitional Phenomena: A Study of the First Not-Me Possession," *The International Journal of Psychoanalysis* 34 (1953): 89–97; and *Playing and Reality* (New York: Routledge, 1989 [1971].

8. On the power of the liminal, see Victor Turner, *The Ritual Process: Structure and Anti-Structure* (Chicago: Aldine, 1969).

9. In the psychodynamic traditions of psychotherapy—inspired by psychoanalytic thought although no longer holding to its orthodoxy—the notion that a personal therapeutic experience is prologue to clinical practice remains strong. Clifford Geertz comments on the relationship between anthropology and depth psychology: "In the study of culture the signifiers are not symptoms or clusters of symptoms, but symbolic acts or clusters of symbolic acts, and the aim is not therapy but the analysis of social discourse. But the way in which theory is used—to ferret out the unapparent import of things—is the same." See Geertz, "Thick Description," 26.

10. See Joseph Smith, *Arguing with Lacan: Ego Psychology and Language* (New Haven: Yale University Press, 1991), 64.

11. In France, for a long time, the intellectual links between interpretive social science and psychoanalysis were part of a general cultural understanding. It was common for young social scientists to undertake what was referred to as a "tranche d'analyse," a slice of analysis. See Turkle, *Psychoanalytic Politics.*

12. William James, *The Varieties of Religious Experience* (New York: Signet, 2003 [1902]), and *The Works of William James: The Principles of Psychology* (Cambridge, Mass.: Harvard University Press, 1981 [1890]). The integration of thought and feeling in our relationships with objects leads us to relate to them with very different styles. On emotionally charged styles of mastery with computational

objects, see Sherry Turkle and Seymour Papert, "Epistemological Pluralism: Styles and Voices within Computer Culture," *Signs: Journal of Women in Culture and Society* 16, no. 1 (1990): 128–157; and Turkle, *The Second Self: Computers and the Human Spirit* (Cambridge, Mass.: MIT Press, 2005 [1984]), particularly chapter 3.

13. Technological objects are not only projective screens over a lifetime, they play an active role in development. So, for example, when children are as young as three, they use objects—the movements of a wind-up toy, the turning of gears—to think through basic questions about aliveness, space, number, causality, and category, questions to which childhood must give a response. See Jean Piaget, *The Child's Conception of the World,* trans. Joan and Andrew Tomlinson (Totowa, N.J.: Littlefield, Adams, 1960 [1929]). At around eight, children's thoughts turn toward winning; they use objects to prove themselves and their ability to control the world. With adolescence, whether children are playing with blocks, LEGOs, dolls, or computers, they move from being focused on control of objects to seeing themselves as builders and designers. Earlier concerns about mastery turn into considerations of identity. Adolescents may be drawn to experimenting with identity in virtual spaces (through a personal Web site, through an avatar); they develop opinions about technology as part of the process of defining who they are.

Metaphysics, mastery, and identity are on the one hand developmental stages and on the other they are lifelong styles of engagement with objects. For more on technology and stages of development, see Turkle, *The Second Self,* part I.

14. Michel Foucault, *Discipline and Punish: The Birth of the Prison,* trans. Alan Sheridan (New York: Pantheon, 1977 [1975]).

15. Verlager makes explicit reference to the work of N. Katherine Hayles. See *How We Became Posthhuman: Virtual Bodies in Cybernetics, Literature, and Informatics* (Chicago: University of Chicago, 1999), 3.

16. Objects take their place in the mourning process. In that process, the things and people that we have lost are brought within the confines of the self. See Sigmund Freud, "Mourning and Melancholia," in *The Standard Edition of the Complete Psychological Works of Sigmund Freud,* trans. and ed. James Strachey, et al. (London: Hogarth Press and the Institute of Psychoanalysis, 1953–1974), vol. XIV, 239–258.

17. In these days of plasticized bodies, humanoid robots, and virtual realities, the quality that Freud called the uncanny, being "known of old, yet unfamiliar," is increasingly important to the study of people's relationships to technology. See Sherry Turkle, *Evocative Objects: Things We Think With* (Cambridge, Mass.: MIT Press, 2007), for memoir fragments on objects that fit this description, most notably a curator's relationship with a mummy.

18. Geertz, "Thick Description," 12.

19. Daniel Bell, *The Coming of Post-Industrial Society: A Venture in Social Forecasting* (New York: Basic Books, 1973).

20. Schüll suggests that one way to think about this is to focus on the stresses of living with what sociologist Barry Schwartz has called an "oppressive abundance of choice." Barry Schwartz, *The Paradox of Choice: Why More Is Less* (New York: ECCO, 2004), 44.

21. Sherry Turkle, *Life on the Screen: Identity in the Age of the Internet* (New York: Simon & Schuster, 1995).

22. On an object feeling continuous with self, see D. W. Winnicott on the "transitional object" in *Playing and Reality.* See also Christopher Bollas, *The Shadow of the Object: Psychoanalysis of the Unthought Known* (New York: Columbia University Press, 1987), 31.

23. Erik H. Erikson, *Childhood and Society* (New York: W. W. Norton, 1963 [1950]).

24. Ibid., 222.

25. Online avatars deployed in virtual communities can embody aspects of self. Looking at one's avatars in a spirit of self-reflection is a powerful new tool for exploring identity.

For an extended discussion of identity and online life, see Turkle, *Life on the Screen.*

26. Christopher Bollas, "The Transformational Object," *The International Journal of Psychoanalysis* 60, no. 1 (1979): 97–107.

27. Lillian Hellman, *Pentimento* (New York: Little Brown, 1973), 1.

28. C. Gordon Bell and Jim Gemmell, "A Digital Life," *Scientific American* 296, no. 3 (March 2007): 63. http://sciam .com/print_version.cfm?articleID=CC50D7BF-E7F 2-99DF-34DA5FF0B0A22B50 (accessed August 7, 2007).

29. Ibid., 58, 60.

30. Susan Yee, "The Archive," in *Evocative Objects: Things We Think With,* ed. Sherry Turkle (Cambridge, Mass.: MIT Press, 2007), 32–36.

31. Susan Sontag, *On Photography* (New York: Dell, 1978), 9.

32. Bell and Gemmell discuss the burdens of having a digital shadow. They anticipate that other people captured by the SenseCam may need to be pixellated so as not to invade their privacy; data would have to be stored "offshore" to protect it from loss and/or illegal seizure; there is danger from "identity thieves, gossipmongers, or an authoritarian state." Bell and Gemmell admit that despite all problems, "for us the excitement far outweighs the fear." Bell and Gemmell, "A Digital Life," 64, 65.

33. Ibid., 63.

34. Ibid., 65.

35. For more on Barry, see Turkle, *The Second Self,* 158–159.

36. Emmanuel Levinas, "Ethics and the Face," *Totality and Infinity: An Essay on Exteriority,* trans. Alphonso Lingis (Pittsburgh: Duquesne University Press, 1969).

The Prosthetic Eye | *Alicia Kestrell Verlager*

1. This is what N. Katherine Hayles refers to as "attempts to transgress and reinforce the boundaries of the subject."

See *How We Became Posthuman: Virtual Bodies in Cybernetics, Literature, and Informatics* (Chicago: University of Chicago, 1999), xiii.

2. Ibid., 3.

3. I first became aware of this way of thinking about my experience in conversation with Sherry Turkle. See Turkle, "Tethering," in *Sensorium: Embodied Experience, Technology, and Contemporary Art,* ed. Caroline A. Jones (New York: Zone, 2006).

Television | *Orit Kuritsky-Fox*

1. Its given Hebrew name, *Ksim Ksam,* is somewhat more poetic and evokes magic and marvel. Once I began to look for its name, I found it quickly, online. Israelis flock to the Internet to discuss their early television experiences. Their online discussions are nostalgic in tone, nostalgic for a time when there was only one Israeli TV channel, public and in black and white. (I was shocked to see color screen shots of *Vision On* when I visited its official Web site.) In the following years, when expensive color TVs would be imported to Israel, the Broadcasting Authority, in the name of equal opportunity, would actually erase the color from the British and American TV shows it aired. (This brought about the soaring sales of even more expensive television sets that came with a canny device called "anti eraser.")

2. Jean Baudrillard, *For a Critique of the Political Economy of the Sign,* trans. Charles Levin (St. Louis: Telos Press, 1981), 54.

3. Interestingly, the ultraorthodox have given the Internet a much warmer welcome. (One of my cousins, a teacher in an ultraorthodox learning center, develops teaching materials for the Internet. The materials are to be used only in the supervised environment of the learning center; computers, like all representational artwork, are forbidden in homes).

The World Wide Web | John Hamilton

1. See Sherry Turkle, "Whither Psychoanalysis in Computer Culture," *Psychoanalytic Psychology: Journal of the Division of Psychoanalysis* 21, no. 1 (Winter 2004): 16–30.
2. In our culture, the association between adolescent males and superheroes is almost universal. For its classic statement, see Joseph Campbell, *The Hero with a Thousand Faces,* 2d ed. (Princeton, N.J.: Princeton University Press, 1968). I use pseudonyms for all patients.
3. It is significant that increased use of weight training and variable resistance forms of physical conditioning have in fact raised boys' expectations of their own bodies. This facilitates the identification with the hypermale figures that they find in toys and online. See Harrison G. Pope et al., "Evolving Ideas of Male Body Image as Seen through Actions Toys," *International Journal of Eating Disorders* 26, no.1 (1999): 65–72.
4. See Stuart T. Hauser and Andrew W. Safyer, "Ego Development and Adolescent Emotions," *Journal of Research on Adolescence* 4, no. 4 (1994): 487–502.
5. There is a site called "Bored of the Rings," which mocks the heroic figures of the Tolkien legend, and there is www .techtv.com, which Eric describes as "morbid humor, like laughing at murders." Eric also mentions www .newgrounds.com.
6. For example, Sherry Turkle, *The Second Self: Computers and the Human Spirit* (Cambridge, Mass.: MIT Press, 2005 [1984]); and *Life on the Screen: Identity in the Age of the Internet* (New York: Simon & Schuster, 1995).
7. Judy Y. Chu, "A Relational Perspective on Adolescent Boys' Identity Development," in *Adolescent Boys: Exploring Diverse Cultures in Boyhood,* ed. Niobe Way and Judy Chu (New York: New York University Press, 2004), 78–104.
8. Sometimes I think that Eric sees me as imaginary as he looks back and forth between me and the screen of imaginary characters. Perhaps I am "in between," in Winni-

cott's sense of in-between playing and reality. See D. W. Winnicott, *Playing and Reality* (New York: Routledge, 1989 [1971]).

9. http://groups.msn.com/AnimeloverHighschool (accessed July 16, 2006).

10. Heinz Kohut, *How Does Analysis Cure?*, ed. Arnold Goldberg, with the collaboration of Paul Stepansky (Chicago: University of Chicago Press, 1984).

Computer Games | *Marsha H. Levy-Warren*

1. Erik H. Erikson, *Toys and Reasons: Stages in the Ritualization of Experience* (New York: W. W. Norton, 1977), 41–43. I use pseudonyms for all patients.

2. I use the term "sense of self" as a way of describing the mental representations that constitute each individual's experience of "I." See Edith Jacobson, "Adolescent Moods and the Remodeling of Psychic Structures in Adolescence," *Psychoanalytic Study of the Child* 16 (1961):164–183; Marsha H. Levy-Warren, *The Adolescent Journey: Development, Identity Formation, and Psychotherapy* (Northvale, N.J.: Jason Aronson, 1996). For a discussion of online selves and their integration into the physical real, see Sherry Turkle, *Life on the Screen: Identity in the Age of the Internet* (New York, Simon & Schuster, 1995).

3. D. W. Winnicott, *Playing and Reality* (New York: Routledge, 1989 [1971]).

4. Marsha H. Levy-Warren, "The Adolescent Journey—I Am, You Are, and So Are We: A Current Perspective on Adolescent Separation/Individuation Theory," *Adolescent Psychiatry* 24 (1999): 3–24; Marsha H. Levy-Warren and Anna K. Levy-Warren, "I am/Might be/Am Not My Diagnosis: A Look at the Use and Misuse of Diagnoses in Adolescence," *Journal of Infant, Child, and Adolescent Psychotherapy* 4, no. 3 (2005): 282–295.

5. Erikson, *Toys and Reasons,* 41–43.

6. A new outcome may follow because the individual has matured since the trauma or because the repetition in

play is happening in a therapeutic context. What was once overwhelming no longer feels that way. Therapy creates an environment for safe expression.

7. The online self becomes an evocative object, an object for self-reflection. See Sherry Turkle, ed., *Evocative Objects: Things We Think With* (Cambridge, Mass.: MIT Press, 2007); and *The Second Self: Computers and the Human Spirit* (Cambridge, Mass.: MIT Press, 2005 [1984]).

8. In Loewald's terms, she "vented" and "sublimated" her feelings. See Hans W. Loewald, *Sublimation: Inquiries into Theoretical Psychoanalysis* (New Haven, Conn.: Yale University Press, 1988).

9. On the ego ideal, see Peter Blos, "The Functions of the Ego Ideal in Adolescence," *The Psychoanalytic Study of the Child* 27 (1972): 93–97; and "The Genealogy of the Ego Ideal," *The Psychoanalytic Study of the Child* 29 (1974): 43–88.

10. With some adolescents, the attunement to their own bodies and to their own fantasy life can become so exquisite that a venture into the real world of clumsy sex is disappointing. In this way, games can mirror the risk-free pleasures of adolescent masturbation.

11. See Loewald, *Sublimation.*

12. Adolescents tend to act before they think, a pattern that relates as much to hormonal shifts as the evolution of cognitive development. It is hard for them to speculate or hypothesize about themselves. See Levy-Warren, *The Adolescent Journey.* They often look at their actions as a way to get more clarity on who they are. Psychotherapy with this age group needs to look at what adolescents do and see it as a statement about who they might be, in order to ask them whether this is who they actually *want* to be. Later, adolescents will be able to think about who they want to be and what they want to do, and whether their actions are in harmony with their goals. See Jean Piaget, "Intellectual Evaluation from Adolescence into Adulthood," *Human Development* 15 (1972):1–12.

13. This is analogous to what I found in my study of young children from chaotic, underprivileged backgrounds. Marsha H. Levy-Warren, "A Child's Play Amidst Chaos," *American Imago* 51, no. 3 (1994): 359–368.

Cyberplaces | *Kimberlyn Leary*

1. This essay is adapted from "Psychoanalytic Selves in A Digital World," in *Postmodern Psychoanalysis Observed,* ed. Joseph Reppen, Jane Tucker, and Martin Schulman (London: Open Gate Press, 2004). The author would like to thank Jonathan Metzl, Daniel Shapiro, and Richard Hale Shaw for their helpful comments on an earlier draft of this manuscript. I use pseudonyms for all patients.

2. Sherry Turkle, *Life on the Screen: Identity in the Age of the Internet* (New York: Simon & Schuster, 1995).

3. See Stephen A. Mitchell and Margaret J. Black, *Freud and Beyond: A History of Modern Psychoanalytic Thought* (New York: Basic Books, 2005); and Sarah Schoen, "What Comes to Mind: A Multiple Case of Patients; Experiences of Good-enough Outcome in Long-term Psychodynamic Psychotherapies." Unpublished doctoral dissertation, University of Michigan, 2001.

4. John Lindon, "Gratification and Provision in Psychoanalysis: Should We Get Rid of 'the Rule of Abstinence'?" *Psychoanalytic Dialogues* 4 (1994): 549–582.

5. Owen Renik, "Getting Real in Analysis," *Psychoanalytic Quarterly* 67 (1998): 566–593.

6. The introduction of Web-enabled technologies to the professional organizations of psychoanalysis has disrupted hierarchies of power, permitting individual members outside the political structures of these organizations an expressive voice to shape debate about the representation of their public identities as psychoanalysts.

   At this writing, correspondents on an Internet-based discussion are actively debating the process by which American Psychoanalytic Association certifies and provides credentials to its members.

7. Turkle, *Life on the Screen.*

8. But how could Melissa know for sure that her knight was only fifteen? She could not. Turkle makes this point in *Life on the Screen.* It is also made by George Johnson in "Lost in Cyberspace: If You Can't Touch It, Can You Steal It?" *New York Times,* December 16, 2001. The knight's "true" identity was endlessly manipulable in cyberspace. But in *presenting* himself as fifteen—whether he was or was not fifteen—Melissa's knight was now perceived as unavailable for the off-screen relationship she desired. She recognized him as someone other than who she had in mind. This functioned as a relational fact, interfering with Melissa's ability to return him to the figure he had occupied in her imagination. I am indebted to Robert Hatcher for raising this question in this context.

9. Sherry Turkle, "Artificial Intelligence and Psychoanalysis: A New Alliance," *Daedalus* 117, no. 1 (Winter 1988): 241–267. See also Turkle, *Life on the Screen*; and "Whither Psychoanalysis in Computer Culture," *Psychoanalytic Psychology: Journal of the Division of Psychoanalysis* 21, no. 1 (Winter 2004): 16–30.

10. Irwin Z. Hoffman, *Ritual and Spontaneity in Psychoanalysis: A Dialectical-Constructivist View* (Northvale, N.J.: Analytic Press, 1998).

11. Owen Renik, "Analytic Interaction: Conceptualizing Technique in the Light of the Analyst's Irreducible Subjectivity," *Psychoanalytic Quarterly* 62 (1993): 553–571; "Publication of Clinical Facts," *International Journal of Psychoanalysis* 75 (1994): 1245–1250; "The Ideal of the Anonymous Analyst and the Problem of Self-disclosure," *Psychoanalytic Quarterly* 3 (1995): 466–495; "The Analyst's Subjectivity and the Analyst's Objectivity," *International Journal of Psychoanalysis* 79 (1998): 487–498; and "Playing One's Cards Face Up in Analysis: An Approach to the Problem of Self-Disclosure," *Psychoanalytic Quarterly* 68 (1999): 521–540.

12. See Owen Renik, "Defining the Goals of a Clinical Psychoanalysis," *Psychoanalytic Quarterly* 71 (2002): 117–124.

13. These ideas remain controversial within psychoanalysis. See Jay Greenberg, "The Analyst's Participation: A New Look," *Journal of the American Psychoanalytic Association* 49 (2001): 359–380. Greenberg, for example, has gone on record as suggesting that psychoanalysis has grown as a discipline under "the impact of one inspired theoretical excess after another" (359), with each new theory focusing narrowly on only a partial truth. He takes particular issue with the "interpersonalization of psychoanalysis"— the tendency of contemporary analysts like Hoffman and Renik to present case reports in which they offer themselves *as persons* to contain the tensions and anxieties patients experience in consequence of being in treatment (364). In so doing, Greenberg suggests that the current focus on mutual influences between patient and analyst unwittingly functions as a prescriptive story for psychoanalysis, every bit as limiting as the traditional authority that the analyst uncritically assumed.

14. See Lindon, "Gratification and Provision in Psychoanalysis."

15. Stephen A. Mitchell, *Hope and Dread in Psychoanalysis* (New York: Basic Books, 1993).

16. Glen O. Gabbard, "Cyberpassion: E-rotic Transference on the Internet," *Psychoanalytic Quarterly* 70 (1997): 719–738. There is the suggestion that psychoanalysis and communication in cyberspace share common potentials as well as common dangers. Each may be deployed as substitutes for actual engagements, even as each may be used at any time to expand the boundaries of self and relation.

The Internal Cardiac Defibrillator | *Anne Pollock*

1. Medtronic, Emergency Room in Your Chest, http://www .medtronic.com/Newsroom/MediaKits.do?category =category.specialty&subcategoryId=1&expand=yes&lang =en_US (accessed May 23, 2006).

2. Eugene Crystal and Stuart J. Connolly, "Evolution of the Implantable Cardioverter Defibrillator," *Lancet* 359, no. 9315 (2002).

3. Kenneth Chang, "Study of Heart-Device Defects Puts Makers on the Defensive," *New York Times,* August 15, 2001.

4. I interviewed thirteen people in the winter of 2002–2003, including three in person and ten over the telephone. All were recruited on the Internet. Two were women with ICDs, two were wives of men with ICDs, and nine were men with ICDs. They included individuals who were rural and urban, from all regions of the United States and one in urban Canada, working class and professional, and ranging in age from twenty-five to seventy-two. All quotations in this chapter are from these interviews. In all cases, I use pseudonyms.

5. While being shocked, the person, much like Victor Turner's description of people undergoing a rite of passage, is in a liminal state, betwixt-and-between life and death. In the setting with emergency room and paddles, this experience and social reintegration occur within a social world. With the ICD, this process does not happen in social space but internally. Victor Turner, *The Ritual Process: Structure and Anti-Structure* (Chicago: Aldine, 1969).

6. For a description of ICD patients' pathways through the experience of implantation, see Suzanne Steffan Dickerson, "Redefining Life While Forestalling Death: Living with an Implantable Cardioverter Defibrillator after a Sudden Cardiac Death Experience," *Qualitative Health Research* 12, no. 3 (2002): 360–372. Her description is of a functional series of stages that lead from loss of control to getting on with living, to the transformation of a life.

7. Sigmund Freud, "The Uncanny," in *The Standard Edition of the Complete Psychological Works of Sigmund Freud,* ed. and trans. James Strachey et al. (London: Hogarth, 1953–1974), vol. XVII, 220.

8. Ibid., 236.

9. Ibid., 235.

10. Audre Lorde writes about not opting for a prosthetic breast after mastectomy; she did not want to hide that she had come close to death. For people with ICDs, it is important that their marker is visible only to them and their closest relations. Audre Lorde, *The Cancer Journals*, 2d ed. (San Francisco: Spinsters/Aunt Lute, 1987).

11. Sandeep Jauhar, "Jolts of Anxiety," *New York Times Magazine*, May 5, 2002.

12. Freud, "The Uncanny," 236.

13. Ibid., 242.

14. Michel Foucault, *Discipline and Punish: The Birth of the Prison*, trans. Alan Sheridan (New York: Pantheon, 1977 [1975]).

15. Slavoj Žižek, "Catastrophes Real and Imagined," *In These Times*, February 28, 2003. http://www.inthesetimes .com/comments.php?id=98_0_4_0_M. (accessed May 17, 2003).

16. This seems to be an intensified version of the notion Joseph Dumit has described as a new pharmaceutical grammar: that being asymptomatic doesn't mean that one should not be on several drugs to avert a statistically possible future risk. See Joseph Dumit, *Drugs for Life: Managing Health and Identity through Facts and Pharmaceuticals* (Durham, N.C.: Duke University Press, forthcoming).

17. Elaine Scarry, *The Body in Pain: The Making and Unmaking of the World* (Oxford: Oxford University Press, 1987).

18. Of course, the reductionism of physicians trained to treat a disease or defect rather than a whole person is not new. See, for example, Robert A. Aronowitz, *Making Sense of Illness: Science, Society, and Disease* (Cambridge: Cambridge University Press, 1998).

19. A high-traffic Internet bulletin board has created a highly developed sociality among ICD recipients. The forum, "The Zapper: For Cardioverter-Defibrillator Implant Recipients, Families, and Care Givers," www.zaplife.org, takes the presence of ICDs in the body of its readers as a given.

On the site, shared experiences with how ICD shock has affected lives show experiences on a broad continuum. All patients emphasized that the experience of an ICD goes beyond the shocks to include such things as the fear of getting shocked at inopportune moments and the embarrassment or mortification when this occurs. Reading the postings leads one to recall Erving Goffman's insight that the presentation of self in everyday life is the foundation of social life. The ICD patient is disrupted in his or her ability to project an idealized self, with significant disruption of all social encounters. See *The Presentation of Self in Everyday Life* (Garden City, N.Y.: Doubleday, 1959).

20. Maureen Dowd, "A Tale of Two Fathers," *New York Times,* Sunday, October 12, 2003.

21. Norbert Wiener, *God and Golem, Inc: A Comment on Certain Points Where Cybernetics Impinges on Religion* (Cambridge, Mass.: MIT Press, 1964).

22. Jacques Derrida, *The Gift of Death: Religion and Postmodernism* (Chicago: University of Chicago Press, 1995).

23. Donna Jeanne Haraway, *Simians, Cyborgs, and Women: The Reinvention of Nature* (New York: Routledge, 1991), 149.

The Visible Human | *Rachel Prentice*

1. The image of the woman can be found at http://www.crd .ge.com/esl/cgsp/projects/video/medical/vishuman .html. The site gives six choices of visible human images to view, including a full-body rendering of the woman and a fly-through of the man's skeleton (accessed September 12, 2007).

2. Sigmund Freud, "The Uncanny," in *The Standard Edition of the Complete Psychological Works of Sigmund Freud,* trans. and ed. James Strachey et al. (London: Hogarth Press, 1953–1974), vol. XVII, 220.

3. Lisa Cartwright, "The Visible Man: The Male Criminal Subject as Biomedical Norm," in *The Visible Woman: Imaging Technologies, Gender, and Science,* ed. Paula

Treichler, Lisa Cartwright, and Constance Penley (New York: New York University Press, 1998), 21–43.

4. Ibid., and Catherine Waldby, *The Visible Human Project: Informatic Bodies and Posthuman Medicine* (New York: Routledge, 2000).

5. Dr. Victor Spitzer, 2000. Personal email. The process involved replacing natural fluids with a more diluted embalming fluid than would typically be used for medical preservation.

6. Waldby, *The Visible Human Project,* 96.

7. Indeed, many people with whom I have discussed the Visible Human Project indicate some familiarity with it by saying something like, "Isn't that the one with the slices?"

8. The only full-body reconstruction I used in this study was that of the visible woman because I could not find a good reconstruction of the visible man on the Internet when I began doing interviews. One reconstruction of the man now can be found at http://www.billkatz .com/VHP/images/man_full_surface.jpg (accessed September 12, 2007).

9. On how technologies enfold multiple histories, see Bruno Latour, "Morality and Technology: The End of the Means," trans. Couze Venn, *Theory, Culture & Society* 19, nos. 5/6 (2002): 247–260.

10. Other informants included a schoolteacher, a dentist, a biologist, a poet, an historian, an economist, a social worker, and a computer-imaging expert. All informants have their identities disguised.

11. I myself have seen a dead human body. Once, as a young reporter on the police beat, I went with an officer to the scene of a death. Lying on a bed in a living room was an old man who had died of cancer. His face was oddly pale to my eyes, and I remember thinking how still he was despite all the chaos of family members, coroner, and police around him.

12. The moral separation of persons and things has been particularly strong in the histories of western cultures.

Igor Kopytoff, "The Cultural Biography of Things: Commoditization as Process," in *The Social Life of Things: Commodities in Cultural Perspective,* ed. Arjun Appadurai (Cambridge: Cambridge University Press, 1986), 64–91.

13. Rachel Prentice, "Bodies of Information: Reinventing Bodies and Practice in Medical Education," Massachusetts Institute of Technology, unpublished dissertation, 2004); and Charis Thompson, *Making Parents: The Ontological Choreography of Reproductive Technologies* (Cambridge, Mass.: MIT Press, 2005).

14. Wendell also is concerned whether digital images, even with all their detail, will show the infinite variations of the body if they are based on only two bodies. This lack of variability among computer-generated and computer graphic anatomical images has raised concerns among some critics, who fear that such standardization of bodies might give clinicians the impression that there is one right body. See Lisa Cartwright, "A Cultural Anatomy of the Visible Human Project," in *The Visible Woman,* 37.

15. On this transformation of the body, see Byron Good, *Medicine, Rationality, and Experience: An Anthropological Perspective* (Cambridge: Cambridge University Press, 1994); and Prentice, "Bodies of Information."

16. D. W. Winnicott, *Playing and Reality* (New York: Routledge, 1989 [1971]), 2.

17. Robert Hertz, *Death and the Right Hand,* trans. Rodney and Claudia Needham (Glencoe, Ill.: The Free Press, 1960 [1907]).

18. Victor Turner, *The Ritual Process: Structure and Anti-Structure* (Chicago: Aldine, 1969).

19. Sherry Turkle, *The Second Self: Computers and the Human Spirit* (Cambridge, Mass.: MIT Press, 2005 [1984]).

20. Margaret Lock, *Twice Dead: Organ Transplants and the Reinvention of Death* (Berkeley: University of California Press, 2002).

21. Sigmund Freud, "Jokes and their Relation to the Unconscious," *The Standard Edition of the Complete Psychological Works of Sigmund Freud,* vol. VIII.

22. Julia Kristeva, *Powers of Horror: An Essay on Abjection,* trans. Leon S. Roudiez (New York: Columbia University Press, 1982).

Slashdot.org | *Anita Say Chan*

1. This and all users' names have been changed for the purposes of anonymity. All interviews were conducted between November 2001 and December 2002.
2. Eve Sedgwick, "Epidemics of the Will," in *Incorporations,* ed. Jonathan Crary and Sanford Kwinter (New York: Zone, 1992).
3. Ibid., 583.
4. On his Web page, Joseph also refers to the practice at Slashdot of users vying to make the first "comment" on a story; he says of his cell phone program: "Also a great tool for first-posters, too ;-)"
5. Slashdot.org. (2002, June 11). UK Government Expands Spying Powers. http://yro.slashdot.org/article.pl?sid =02/06/11/156240&mode=nested&tid=158. (accessed December 10, 2002). The original bill had previously limited online surveillance powers to a select body of law enforcement and intelligence agencies, but an expanded bill would permit several hundred local and national agencies such access. Among the comments posted to the conversation were sample letters written by users against the bill, a link to the fax numbers of Parliament members to contact by phone to register opposition, and a summary of a U.S. Senate Committee's 1976 review of intelligence activity in the Nixon administration as an example of unrestricted government surveillance.
6. The drug user had been seen as inhabiting a predictable state of "relative homeostatic stability and control" that was maintained by substance dependence; the drug addict was reframed in a narrative of uncontrolled self-decline that required a vast and unbounded system of treatment. Whether or not the institutional disciplines that now came to surround addicts can help, these institutions nevertheless "presume to know her better than

she can know herself—and indeed, offer everyone in her culture who is *not herself* the opportunity of enjoying the same flattering presumption." Sedgwick, "Epidemics of the Will," 582.

7. David N. Greenfield, *Virtual Addiction: Help for Netheads, Cyberfreaks, and Those Who Love Them* (Oakland, Calif.: New Harbinger Publications, 1999); and Kimberly S. Young, *Caught in the Net: How to Recognize the Signs of an Internet Addiction—and a Winning Strategy for Recovery* (New York: John Wiley, 1998). Writing on Internet addiction includes an explicitly Christian reformist interpretation of the phenomenon. Books by Andrew Carega and Steven O. Watters attempt to convince their readers of the sinfulness of overindulging in the pleasures and decadences of the Internet and, not surprisingly, point their readers toward a list of Christian Recovery Groups to help redeem the fallen net user. See Andrew Careaga, *Hooked on the Net: How to Say "Goodnight" When the Party Never Ends* (Grand Rapids, Mich.: Kregel Publications, 2002); Steven O. Watters, *Real Solutions for Overcoming Internet Addictions* (Ann Arbor, Mich.: Vine Books, 2001).

8. See Joseph Weizenbaum, *Computer Power and Human Reason: From Judgment to Calculation* (San Francisco: W. H. Freeman, 1976), 116.

9. Ibid.

10. Ibid.

11. Ibid., 122.

12. Paul Taylor, *Hackers: Crime in the Digital Sublime* (London: Routledge, 1999), xii, 116.

13. Andrew Ross, "Hacking Away at the Counterculture," in *Technoculture: Cultural Politics,* ed. Constance Penley and Andrew Ross (Minneapolis: University of Minnesota Press, 1991), 120.

14. Mary Douglas, *Purity and Danger: An Analysis of the Concepts of Pollution and Taboo* (New York, Routledge, 1966).

15. Ibid., 36.

16. See Foucault, *Discipline and Punish: The Birth of the Prison*, trans. Alan Sheridan (New York: Pantheon, 1977 [1975]).

17. Slashdot.org. (2002, March 1). Announcing Slashdot Subscriptions. http://slashdot.org/article.pl?sid=02/ 03/01/1352200&mode=thread&tid=124. (accessed December 10, 2002).

18. One said, "I'm what a proper Slashdot 'citizen' should be. I'm one of the many people who keep this site from falling into the swamp of the trolls. By rights, [the editors] should be *paying* me. I'm practically a Slashdot employee, and me and people like me are what brings Slashdot its value. No good citizens, no Slashdot."

19. The notion of addiction on Slashdot is the product of boundary work. At stake are how such terms as *normalcy* and *excess*, and *productivity* and *waste* are defined. Thomas F. Gieryn writes eloquently on this topic. See "Boundary-work and the Demarcation of Science from Non-science: Strains and Interests in Professional Ideologies of Scientists," *American Sociological Review* 48 (1983): 781–795; "Balancing Acts: Science, Enola Gay, and History Wars at the Smithsonian," in *The Politics of Display,* ed. Sharon Macdonald (London: Routledge, 1999); and *Cultural Boundaries of Science: Credibility on the Line.* (Chicago: University of Chicago Press, 1999).

The Dialysis Machine | *Aslihan Sanal*

1. Jacques Lacan, *Ecrits:A Selection,* trans. Alan Sheridan (New York: W. W. Norton, 1977), and *The Language of the Self: The Function of Language in Psychoanalysis,* trans. Anthony Wilden (Baltimore: Johns Hopkins University Press, 1968).

2. The identities of all informants have been disguised.

3. Virginia Woolf, *To the Lighthouse* (New York: Harcourt, Brace, 2005), 65.

4. Turkish scholars believe that the Turks' original religion was a form of shamanism that later mingled with Islam. Scholars note commonalities between shamanism and

local Muslim saint-worship practices. These commonalities are discussed in studies of early Turkish texts such as the Orhun and Urgur tablets and the oral tradition of Dede Korkut, stories spread throughout Turkic communities in Asia and Anatolia. See Ziya Gökalp, *The Principles of Turkism,* trans. Robert Devereux (Leiden, Netherlands: E. J. Brill, 1968); and Bahaeddin Ogel, *Islamiyetten Once Turk Kultur Tarihi* (Orta Asya kaynak ve buluntularina gore) (Ankara: Turk Tarih Kurumu Basimevi, 1991).

5. On one hand, a *dede* is respected for his political and worldly role; on the other, he is viewed as a spiritual leader. Unlike shamans, *dedes* do not conduct exorcism or treat patients. Their role is social: to unite the community, to open and close prayer sessions with poems, sing gospels, and play the *saz. Dedes* speak about sociopolitical issues; they educate the youth on Alevi culture and make peace among rivals. Cemal Sener, *Alevi törenleri: Abdal Musa, Veli Baba Sultan, Hamza Baba, Haci Bektas Veli.* Çemberlitas (Istanbul: Ant Yayinlari, 1991).

6. Mary Douglas, *Purity and Danger: An Analysis of Concepts of Pollution and Taboo* (New York: Routledge, 1966).

7. Hüseyin Bal, *Alevi-Bektasi köylerinde toplumsal kurumlar* (Istanbul: Ant Yayinlari, 1997).

8. Victor Turner, *The Forest of Symbols: Aspects of Ndembu Ritual* (Ithaca, N.Y.: Cornell University Press, 1967), and *The Ritual Process: Structure and Anti-Structure* (Chicago: Aldine, 1969).

9. In his essay "Mourning and Melancholia," Freud argues that a loss of a loved one damages the ego. With time, the lost object can be internalized in the work of reparation known as mourning. Without successful internalization, the unfilled emptiness may take the shape of melancholia. A dialysis patient is caught in terms resonant with Freud's narrative, terms of emptiness and being filled. Oguz is shaken by an experience of being emptied out (recalling loss) and being filled in (which should be part of a recovery from loss). But unlike psychological mechanisms that cure through provision, dialysis as a medical

solution tries to cure through a process, both invasive and recurrent, that some patients experience as depletion. This is how Oguz experienced it, sending him on a dangerous quest for satiation. Oguz, in an attempt to get rid of emptiness, drinks more than he is supposed to, which causes damage to his body. See Sigmund Freud, "Mourning and Melancholia," in *The Standard Edition of the Complete Psychological Works of Sigmund Freud,* ed. James Strachey, et al. (London: Hogarth Press and the Institute of Psychoanalysis, 1953–1974), vol. XIV.

10. His doctor shared the same information with me; he told me that in his transplant unit, he had never had a patient who lived with a new kidney for more than nine years.

11. See Victor Tausk, "The Influencing Machine," in *The Psychoanalytic Reader: An Anthology of Essential Papers, with Critical Introductions,* ed. Robert Fliess (New York: International Universities Press, 1992).

Video Poker | *Natasha Schüll*

1. This essay draws from a book-length ethnographic study by the author based on fieldwork in Las Vegas among gamblers and technology designers in the gaming industry. The identities of all my informants have been disguised.

2. Roger Caillois, *Man, Play, and Games* (New York: Free Press of Glencoe, 1961).

3. Erving Goffman, *Where the Action Is: Three Essays* (London: Allen Lane, 1969).

4. Clifford Geertz, "Deep Play: Notes on the Balinese Cockfight," in *The Interpretation of Cultures* (New York: Basic Books, 2000 [1973]).

5. Goffman, *Where the Action Is,* 270.

6. Geertz, "Deep Play," 424, 436.

7. A recent upswing in the popularity of live and online poker is directly related to the television broadcast of the 2003 World Series of Poker, during which an amateur won the $2.5 million top prize. Until then, poker was regarded as

a waste of casino floor space in the gaming industry; as a skill-based game played between individuals rather than against the house, there is no way for casinos to have an edge (instead, they take small buy-ins from players). The few establishments that kept open poker rooms did so to attract wealthy clientele who wished to play against the champions. Since poker went on TV and became popular, more casinos offer the activity. Yet even with live poker currently at the height of its popularity, a mere 3 percent of consumers identify it as their preferred game while 61 percent choose slot machines (American Gaming Association, *State of the States: The* AGA *Survey of Casino Entertainment*, 2007, 35). A 37 percent increase in poker revenue in 2005 slowed to 15 percent in 2006; it remains to be seen whether this will continue to decline, or plateau. A Las Vegas investment analyst predicts: "Poker may never be a money-maker like slot machines, blackjack, baccarat, and craps, but it has its place now" (Dave Ehlers, quoted in Jeff Simpson, "Casinos Bet on Poker's Popularity," *Las Vegas Sun,* April 2, 2004. http:// www.casinocitytimes.com/news/article.cfm?contentID =141985) (accessed on September 12, 2007).

8. Robert Hunter, personal communication, March 1999.

9. On standard video poker machine consoles, players press a DEAL button whereupon five cards appear on the screen each with a HOLD button beneath it. Players choose which cards they wish to keep and press the DEAL button again, which renders new cards to replace those discarded. Typically, a player has the option to bet one to five credits, valued from twenty-five cents to five dollars.

10. Robert Hunter, personal communication, March 1999.

11. For this traditional analysis, see Nerilee Hing and Helen Breen, "Profiling Lady Luck: An Empirical Study of Gambling and Problem Gambling amongst Female Club Members," *Journal of Gambling Studies* 17, no. 1 (2001): 47–69; J. Koza, "Who Is Playing What: A Demographic Study (part 1), *Public Gaming Magazine* (1984); Henry R. Lesieur and Sheila B. Blume, "When Lady Luck Loses: Women and Compulsive Gambling," in *Feminist Perspec-*

tives on Addictions, ed, N. Van Den Bergh (New York: Springer, 1991); Susan D. McLaughlin, "Gender Differences in Disordered Gambling," *National Council on Problem Gambling* (2000). More historically minded explanations might understand the gender skew as an artifact of women's culturally framed relationship to domestic appliances and specifically to slot machines—long regarded as "toys" to occupy them while their male companions played table games.

12. Having begun his book with abject scenes of gamblers wearing adult diapers and urinating into cups next to their machines, Jackson Lears clarifies his intentions: "I do not mean to dignify compulsive gambling or deny its damaging effects." See Jackson Lears, *Something for Nothing: Luck in America* (New York: Viking Press, 2003), 10.

13. Elsewhere, I examine the strategies by which game developers engineer gambling machines to facilitate self-escape. See Natasha Schüll, "Digital Gambling: The Coincidence of Desire and Design: Cultural Production in a Digital Age," *Annals of the American Academy of Political and Social Science* 597 (2005): 65–81.

14. Nikolas Rose, *Powers of Freedom: Reframing Political Thought* (Cambridge: Cambridge University Press, 1999), 164. See also Rose, *Inventing Ourselves: Psychology, Power, and Personhood* (Cambridge: Cambridge University Press, 1998); and Peter Miller, "Governing by Numbers: Why Calculative Practices Matter," *Solid Research* 68, no. 2 (2001), 379–396.

15. Rose, *Powers of Freedom,* 152.

16. Ibid., 214.

17. Michel Foucault, "Technologies of the Self," *Technologies of the Self: A Seminar with Michel Foucault,* ed. Luther H. Martin, Huck Gutman, and Patrick H. Hutton (Amherst: University of Massachusetts Press, 1988), 16–49; and *The History of Sexuality,* vol. 3, *The Care of the Self* (New York: Vintage, 1990).

18. Pat O'Malley, "Risk and Responsibility," in *Foucault and Political Reason: Liberalism, Neo-Liberalism, and Rationalities of Government,* ed. Andrew Barry, Thomas Osborne,

and Nikolas Rose (Chicago: University of Chicago Press, 1996), 198.

19. Rose, *Inventing Our Selves,* 87. See also both Anthony Giddens, "Living in a Post-Traditional Society," and Ulrich Beck, "The Reinvention of Politics: towards a Theory of Reflexive Modernization," in *Reflexive Modernization: Politics, Tradition, and Aesthetics in the Modern Social Order,* ed. Ulrich Beck, Anthony Giddens, Scott Lash (Stanford, CA: Stanford University Press, 1994).

20. Barry Schwartz, Hazel Rose Markus, and Alana Conner Snibbe, "Is Freedom Just Another Word for Many Things to Buy? That Depends on Your Class Status," *The New York Times Magazine,* February 26th, 2006, 14–15. Alberto Melucci has written that "choosing is the inescapable fate of our time." See *The Playing Self: Person and Meaning in the Planetary Society* (Cambridge: Cambridge University Press, 1996).

21. Barry Schwartz, *The Paradox of Choice: Why More Is Less* (New York: ECCO, 2004), 2.

22. Ibid., 44. See also Edward C. Rosenthal, *The Era of Choice: The Ability to Choose and Its Transformation of Contemporary Life* (Cambridge, Mass.: MIT Press, 2005).

23. Roy F. Baumeister, *Escaping the Self: Alcoholism, Spiritualism, Masochism, and Other Flights from the Burden of Selfhood* (New York: Basic Books, 1991), 211.

24. Daniel Bell, *The Coming of Post-Industrial Society: A Venture in Social Forecasting* (New York: Basic Books, 1973).

25. Mike Davis, "Class Struggle in Oz," *The Grit Beneath the Glitter: Tales from the Real Las Vegas,* ed. H. K. Rothman and M. Davis (Berkeley: University of California Press, 2002).

26. Arlie Hochschild, *The Managed Heart* (Berkeley: University of California, 1983), 5.

27. Ibid., 11.

28. Lears, *Something for Nothing,* 8.

29. Ibid.

30. Goffman, *Where the Action Is,* 261.

31. Sherry Turkle, *The Second Self: Computers and the Human Spirit,* (Cambridge, Mass.: MIT Press, 2005 [1984]), 85.

One of the game players Turkle spoke with told her: "You know what you are supposed to do. There's no external confusion, there's no conflicting goals, there's none of the complexities that the rest of the world is filled with. It's so simple. You either get through this little maze so that the creature doesn't swallow you up or you don't" (84).

32. Sigmund Freud, "Beyond the Pleasure Principle," in *The Standard Edition of the Complete Psychological Works of Sigmund Freud,* trans. and ed. James Strachey, et al. (London: Hogarth Press and the Institute of Psychoanalysis, 1953–1974), vol. XVIII, 7–64.

33. William Bennett, *The Book of Virtues: A Treasury of Great Moral Stories* (New York: Simon & Schuster, 1996). "We should know that too much of anything, even a good thing, may prove to be our undoing. . . . We need . . . to set definite boundaries on our appetites" (53).

34. Joshua Green, "The Bookie of Virtue: William J. Bennett Has Made Millions Lecturing People on Morality—and Blown It on Gambling," in *The Washington Monthly* (June 2003).

Index

Office of National Drug Control Policy, 170
O'Malley, Patrick, 157

Patterning table, 49–54
Perl (computer language), 129–130
Piaget, Jean, 174n13
Play, theories of, 78–79, 179n6
Poe, Edgar Allan, 146, 148
Poker
  television broadcasts of, 193–194n7
  upswing in popularity of, 193–194n7
  video, 14–15, 153–171
Pollock, Anne, 12–14, 98–111
Prentice, Rachel, 17, 112–124
The prepared listener
  the ethnographer as, 7, 8
  in inner history, 6–11
Prosthetic eyes, 12–14, 32–40
Psychoanalysis
  computational models used in, 87
  and cyberspace, 92–95
  credentials for and politics of, 181n6
  defining, 10
  as discipline of self-reflection, 7–11
  emails and the analytic couple, 90–92
  evolution of in cyberspace, 92–95
  providing analytic "gratification and provision" during, 86
  theoretical alliance in, 87
Psychoses, cultural, 144
Purity and Danger (Douglas), 134

Renik, Owen, 92–94
Rett Syndrome, 24, 50–54
Role playing, and the World Wide Web, 66–73
Rorschach, technology as, 11, 174n13. *See also* Styles of
  mastery
Rose, Nikolas, 157
Ross, Andrew, 134

"Safe space" in psychoanalysis and ethnography, 7, 9,
  179–180n6

Transitional objects, 118, 173n7, 175n22. *See also* Winnicott, D. W.
Transitional space, 6, 78–79. *See also* Winnicott, D. W.
Transplants, kidney, 138–152
Turkle, Sherry, 2–29, 67, 87, 168, 177n3, 196–197n31
Turner, Victor, 184n5. *See also* Liminality

U.S. intelligence committee, 189n5
U.S. National Library of Medicine, 113

Verlager, Alicia Kestrell, 12, 14, 20, 32–40
Video poker, 14–15, 153–171
  addiction to, 18
  gender and, 194–195n11
  "machine life" and 167–170, 196–197n31
  money and time and, 161–167
  the self conceived as enterprise and, 156–158
  "self-liquidation" in, 158–161, 169
  time distortion and, 161–167
Violence, in computer games, 82–85
Vision On (TV show), 55–56, 59, 177n1
The Visible Human Project, 16–18, 112–124, 187n4, 187nn6, 7
  Cybil (case study), 118–121
  Julie (case study), 122–124
  Visible Human female, 112, 186n1, 186n3, 187n8, 188n14
  Visible Human male, 186nn1, 3, 187n8
  Wendell (case study), 115–118

Watters, Steven O., 190n7
Weizenbaum, Joseph, 134
Wiener, Norbert, 110
Winnicott, D. W., 78–79, 118, 173n7, 175n22, 178n8
Woolf, Virginia, 139, 141
Working addicts, on Slashdot.org, 130–132
World Wide Web
  blogging, 73–75
  clinical work on, 64–95
  role playing, 66–73

July 9, 2009 Amazon 16.47 107430